Hidden History Beneath Folsom Lake

Hiking Across a Dry Lake Bed in Time of Drought

By Kevin Knauss

Edited by Bonnie Osborn

Published by Kevin Knauss

Dedicated to my son Walker, whose grit and determination in High School made writing this book look easy. – Kevin Knauss

Granite Bay, CA.

www.insuremekevin.com

Contents

Hidden History Beneath Folsom Lake

Hiking Across a Dry Lake Bed in Time of Drought

1952 aerial photograph taken during the construction of Folsom Dam showing the North Fork and the South Fork American Rivers before being covered in lake water.

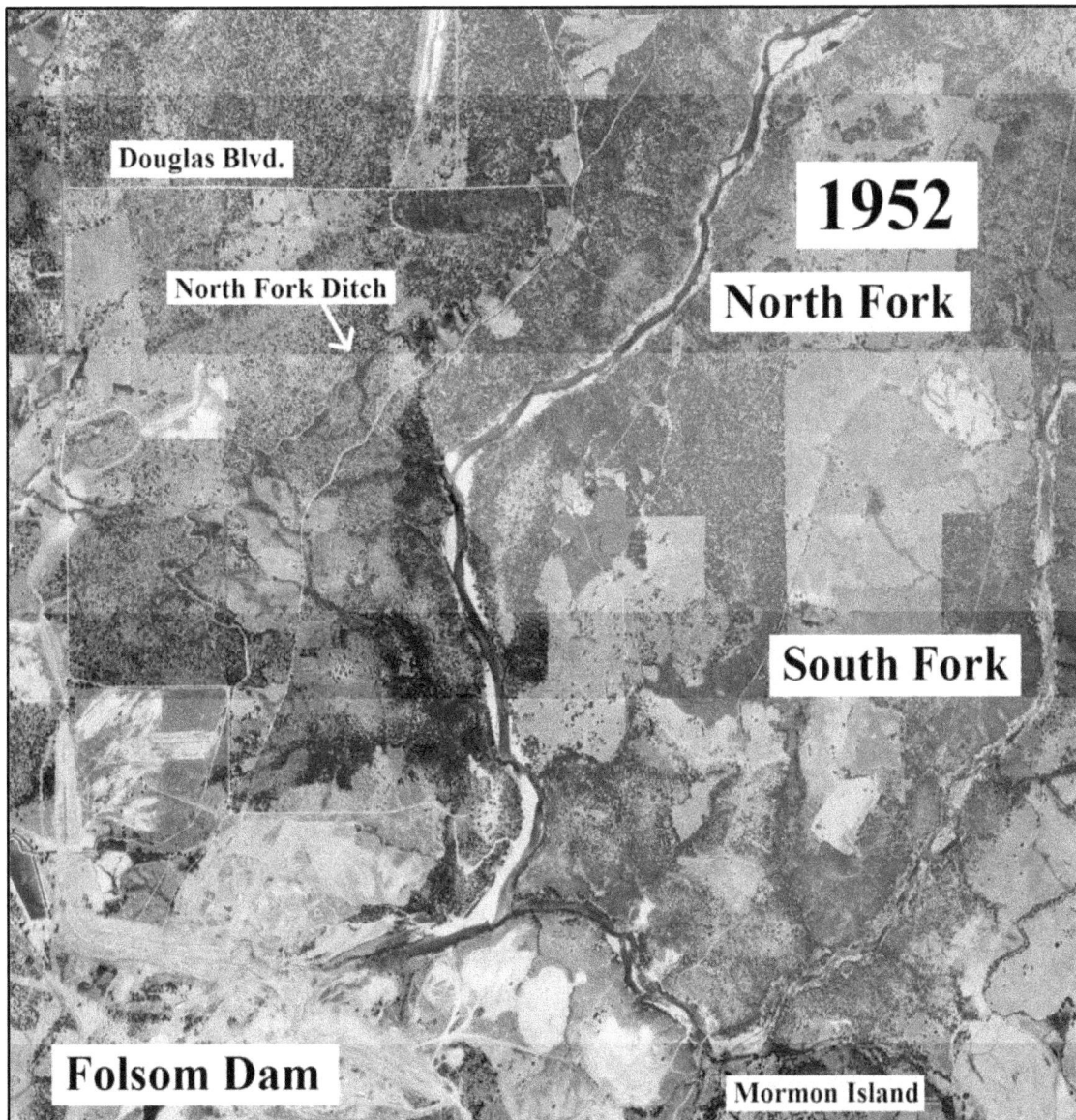

Introduction

Reflections of the 2015 drought at Folsom Lake

I will admit that I was sad to see the rains come and Folsom Lake begin to fill in the autumn of 2015. It had been a glorious summer of hiking over a dry lake bed revealed by the lowest lake level since Folsom reservoir was constructed in the 1950s. While the exposed lakebed was devoid of any real vegetation that might have been present before the lake was filled, it still felt like I was walking into the past. As an amateur historian I'm not immune from fantasizing about going back in time and experiencing the life of the early gold miners, pioneers or Native Americans.

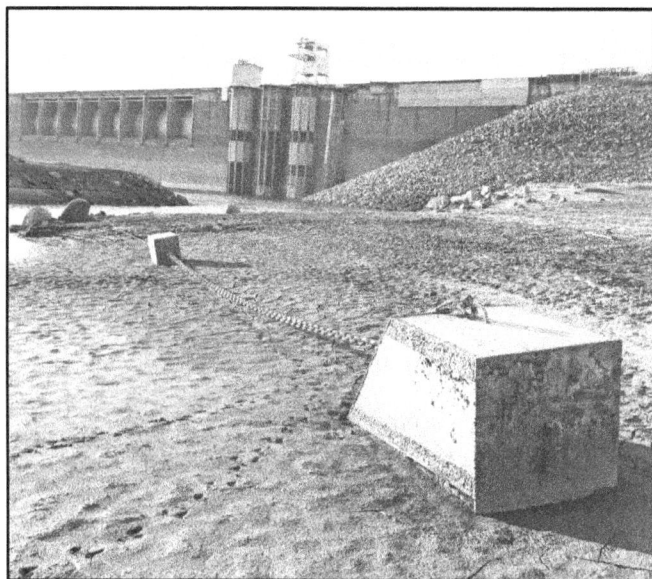

1. November 21, 2015: Folsom Lake was reduced to 15% of capacity with an elevation of approximately 350 feet.

With Folsom lake at historically low levels brought on by drought, coupled with planned and abnormally high releases of reservoir water for environmental reasons, a variety of historical sites were exposed for the first time since the 1950s. A drastically shrunken lake meant that I was able to ford across the North and South forks of the American River in places that were normally covered with lake water. I sloshed across the river with my gear like so many other men and women before me, in search of food, fur or gold, only I was in search of remnants of history.

Over the course of the drought lake levels of 2014 and 2015, I took numerous hikes exploring the dry lake bed that had once been gold mining operations, small communities, and the land of Native American tribes. This book is a compilation of those hikes and others as I explored the Folsom and Auburn State Recreational Areas. My hikes took me from the base of Folsom Dam, up the North Fork to Rattlesnake Bar, and below Auburn. I travelled along most of the South Fork from Mormon Island up to Salmon Falls. Most of these hikes and historical finds were previously documented in blog posts on my website along with photos and images from old maps.

Except for some good rain storms at the end of 2014, the precipitation all but stopped as 2015 began. The region recorded one of the driest Januarys on record in 2015. There was virtually no snow in the Sierras to melt and send much water into Folsom from the North, Middle, or South forks of the American. Then, in the spring of 2015, the Bureau of Reclamation, in cooperation with other dam operators, decided to enhance the survivability of the winter run Chinook salmon. The goal was to retain as much cold water as possible behind Shasta Dam to be released in the fall. With decreased flows from Lake Shasta and Oroville, abnormally high flows of water would be released from Folsom to repel salinity intrusion in the Delta. The Bureau of Reclamation announced that Folsom would be drained to some of its lowest levels in history by the end of August 2015. The lake hit a low elevation level of 348.68 feet on December 4th, or approximately 14% of capacity.

Hidden History Beneath Folsom Lake consolidates my many hikes and photographs across the dry lake bed into one book with additional historical background. Most of the sites are only visible when the reservoir level is low, and many remain hidden and inaccessible until the lake level reaches critical drought stage of less than 20% of capacity. This book arranges my various hikes into a tour narrative based on geographical location. I start on the west side of Folsom Dam, south of Beals Point. I then trace the various historical points of interest counter-clockwise up the North Fork of the American River, down the Peninsula, and finish up traveling up the South Fork to Salmon Falls. Undoubtedly there are items I missed. Invariably, on hikes over ground that I may have walked over only a month before, I would see a new object or topographical feature that I previously had overlooked. Consequently, I'm certain my catalog of the historical sites I visited is not complete.

It is not my intention to provide a full recitation of all the historical places or events that shaped the North and South forks region of the American River that lie beneath Folsom Lake. Rather, I attempt to document what I came across on my various hikes and give a little historical perspective to those remnants of history that remain, usually under the water of Folsom Reservoir.

I endeavored to verify the name, places and history associated with such to the best of my ability. I try not to assume or speculate in a manner that will lead the reader to refer to my text as historical fact. The only verifiable facts are my observations and the photos that I have included in this book. When possible, I've tried to cross reference sites and structures with maps that I've included. However, it is not uncommon for a place or structure to have been referenced with three different names over the course of decades in newspaper articles or on maps.

For the sake of those who may venture out across Folsom at low lake levels, I default to the last title used on the most recent document. For example, the water ditch on the south side of the South Fork of the American River was often referred to as the Natoma Ditch in stories and old maps. I refer to it as the Natomas Ditch because that is the last title attributed to it on the United States Geological Survey (USGS) topographical maps. In addition, because so many of these historical places underneath Folsom Lake are no longer identified on current maps, I also use the current place names as a reference point. For example, I discuss hiking around Dotons Point in the Folsom Lake Recreation Area. Historically this area was generally known as the mining town site of Carrolton.

Rivers don't run in a straight line. The North and South forks of the American River can make odd turns with the topography. I have endeavored to describe locations in reference to maps and a compass. Consequently, I might reference the south side of the North Fork, even though it generally runs north to south, because there is a bend in the river where it is flowing east to west.

When it comes to articles from 19th century newspapers, the only solid facts are that black ink was used on white newsprint. In other words, published articles had frequent errors and embellishments by the author. One series of articles on the North Fork Ditch kept referring to the Bear River as being within close proximity to the American River. It was also a matter of style to promote future mining prospects with a bright and prosperous future. Regardless, many newspaper articles of the time provide important observations about the mining activity along the American Rivers.

I welcome the comments of others who may have more knowledge or reference material to either refute my research or provide additional verifiable details as to the historical places and appurtenances detailed in this book. An invaluable asset to my knowledge base for this book was the research done by John H. Plimpton. His collection of history on the Middle, North and South forks of the American River within the boundaries of Folsom Lake comprises many binders at the Placer County Archives room. Mr. Plimpton worked for the California State Parks Department and did extensive research in the field as well as through public records to catalog the earliest history of the American River.

Another source of wonderful historical material has been gleaned from the California Digital Newspaper Collection, Center for Bibliographic Studies and Research, University of California, Riverside, http://cdnc.ucr.edu. All newspaper quotes are from this digital collection.

Finally, I apologize in advance for some of the small images included in this book. A hard disk crash wiped out many photos I had taken on certain hikes. Consequently, I had to use photographs that had been reduced in image size for inclusion in blogs I had posted to my website.

When these images are converted from 72 dpi (web based) to 300 dpi (for printed material) they necessarily are much smaller. Larger examples of these images can be found in the ebooks of the same name for download to computers and mobile devices.

Geology

The geology of the Folsom Lake area is a mixture of granitic plutons that have been lifted up through an overlay of metamorphic rock. Deposited on top of this mix are volcanic ash and debris flows that came down from higher elevations. Primarily on the North Fork, the American River has cut through the unconsolidated debris, creating steep banks on either side. The South Fork of the American River within Folsom Lake winds around steep hillsides of granitic and metamorphic rock in El Dorado County.

On both forks of the American River, it was the placer gold (small flakes, nuggets, and dust) that the miners were searching for. The highest concentration of placer gold was usually deposited in the river below the smooth river cobble rocks on either the granitic or metamorphic bedrock underneath. Hence, we have the early miners panning for gold in the river and over time developing mining methods to more efficiently remove gold from the river banks. The later mining operations, sometimes by hydraulic methods, have left mounds of rock cobble tailings along and above the original river bed.

The miners assumed that the placer gold wasn't only reserved for the American River of the 1850s, but also might be present in old river beds adjacent to the current river course. This led to several mines where the miners would excavate the debris flow of mud and river cobble on top of granitic plutons. The goal was to dig down to the granite base in search of gold. One such mine was located on a ridge sandwiched between Dyke 5 on the west and Dyke 4 on the east in Granite Bay. The mining pit was located on top of what I will now refer to as Gold Mine Ridge and the mine exit on the north side of the ridge on a horse stable property.

River channel

The channels of the North and South forks of the American River at their confluence had relatively steep banks on either side of the stream. The early miners had to walk down 20 or 40 feet just to get to the gravel bars to pan for gold.

The riverbed elevation at the confluence was approximately 250 feet above sea level before the dam was built. The lowest lake level elevation at the end of November 2015 was 349 feet elevation. The water elevation of a full Folsom Lake is 466 feet. Even with a heavily silted river channel there was still a good 80 feet of water depth to the river channel at the lowest lake level in the autumn of 2015. An exciting event for a hiker and historian like me was being able to witness the American River running free when in normal years the area would be lake water. On the North Fork this occurred at Horseshoe Bar, and on the South Fork the river ran into the lake water below Higgin's Point or the Salmon Falls region.

American River Mining

It was primarily the gold mining industry that created the relics of history beneath Folsom Lake. The mining pioneers first directed their efforts at the river bars on the American River. Situated next to the flowing river, the bars of

sand and gravel could be shoveled into pans and rockers to sift out the gold. As the placer gold of the accessible river bars diminished, the miners turned their attention to the actual river bed. But to get to the gold that lay under a couple feet of flowing water, the river had to be diverted out of its channel.

Daily Alta California, Volume 2, Number 218, 17 July 1851

This stream [North Fork American River] has been falling for a month past about a foot a week. Within the last three days, in consequence of the cold weather experienced on the mountains, the water has not lowered perceptibly; but as the snow has now disappeared, we may look for a gradual and steady fall until the equinoxial storms set in. The river is now but nine inches higher than in September, last year.

Almost the entire stream, from its junction with the Middle Fork down to Mormon Island, will be diverted from its original channel during the ensuing fall. The most extensive operations are at Condemned Bar, Beales' Bar, Willow Bar, Mormon Bar, Kentucky Bar, New York Bar and Oregon Bar. The latter company have completed their canal, and nearly finished their dam, which has been constructed at immense expense, and in the most substantial manner. No ordinary freshet can make the slightest impression upon it.

The mania for damming is as great as that for speculating in quartz veins, and many have commenced their labors, not only at points where there is no evidence that gold exists, but where experiments were made last year which proved utter failures. However, the extreme low stage of the water this year will undoubtedly enable miners to reach those placers, in the bed of the streams, which have ever since the discovery of gold been unacquainted with the insinuating pick of the adventurous miner.

Newspaper articles in the fall of 1851 confirmed that the North and South forks of the American River, below the confluence of the Middle Fork and Salmon Falls respectively, had almost entirely been turned out of its original channel. The river was channeled into flumes or canals in order to dry out the riverbed for gold mining.

After the riverbed had been scoured clean of gold, the miners turned their attention to the gold contained in the placer benches and banks along the river. But to tackle the landscape above the river they would need another source of water to efficiently mine the river banks. The demand for water above the mining claims drove the development of the North Fork Ditch, Natomas Ditch and the Negro Hill Ditch.

Remnants of these ditches or water canals, mainly below the water line of Folsom Lake, constitute many of the hidden historical relics underneath the reservoir. The incredible amount of physical labor to dig and construct these ditches is evident. These water canals represent a monumental undertaking given the challenges faced in the 1850s, from steep slopes, the necessity of chiseling through solid granite, and the lack of mechanized tools.

Water Diversions

Gold mining along the American Rivers would not have lasted long without the advent of water canals. The only way for miners to efficiently tap into the placer gold held in the river banks was to have a source of water higher in elevation. The construction of the water canals or ditches

allowed the miners to run sluice boxes and in some cases hydraulic monitors to wash the earth for gold.

North Fork Ditch

In the summer of 1853 an initial survey was made for the construction of the North Fork Ditch from Auburn down to Beals Bar. The first dam, flumes, and ditch were finished in January 1855. The North Fork Reservoir was originally constructed in 1856.

Sacramento Daily Union, Volume 7, Number 1084, 13 September 1854

North Fork Canal.- This stupendous enterprise, projected but a little over one year since, is already in the full tide of successful prosecution, and will undoubtedly be completed within the ensuing month. The route of the canal from Tahmaroo Bar, near the junction of the Middle, [Fork] to below Beal's Bar, at the junction of the South Fork with the North Fork of the American River, was surveyed in August of last year. The entire distance from the source of the canal to its mouth is twenty-five miles, and the point of disemboguement is but two miles distant from Mormon Island. Although the fall of the river in this distance is great, still so high are the banks, that at Beal's Bar the canal is at least one hundred feet above the river.* The ditch is one of the largest in the country, being ten feet on top, four feet on the bottom, and four feet in depth. The contract has been let to Messrs. Brooks and Clark, who have sub-let the same, and workmen are now busily employed along the entire route.

The mineral district through which this canal traverses is indubitably among the most valuable of any in the rich county of Placer. It traverses a section about midway between the Sacramento and Auburn road, and the North Fork, and will furnish an abundant supply of water to a strip of auriferous territory heretofore valueless, in consequence of the scarcity of this indispensable element to successful mining.

*The NFD Reservoir sat at an elevation of 400 feet. The river bed was at 250 feet elevation for a total difference of 150 feet.

Frequent floods would break apart numerous dams at the headwaters of the ditch and wash away flumes over the ravines. Sometimes the repairs would take months and leave the miners without any water to continue their operations. Over the years the North Fork Ditch was strengthened, fortified and eventually lined with concrete. When Folsom Lake drops low enough, you can see the remnants of this important water delivery system up the North Fork of the American River.

Natomas Ditch

On the South Fork of the American River, water ditches were constructed on both sides of the river. A dam was constructed upstream of the mining camps at Salmon Falls to divert water into the ditches. The ditch on the south side of the river, known as the Natomas Ditch, carried water all the way down to Mormon Island and eventually into the City of Folsom. The Negro Hill Ditch on the north side of the river delivered water down to the mining community of Negro Hill and north up to Massachusetts Flat. The Natomas Ditch was initially completed to Mormon Island in 1853.

Sacramento Daily Union, Volume 5, Number 658, 3 May 1853

New Diggings. — We understand from a gentleman residing near Mormon Island, that new diggings of great extent have just been

discovered between that place and McDowell Hill, yielding an average of from three to five cents to the bucket. Some of the prospects have been as high as 50 and 60 cents. The canal of the Natoma Company has been completed within a few days to Mormon Island, and is now supplying that whole section with water, at which point miners are congregating from all quarters. We are also informed that the Natoma Company are extending their canal as rapidly as possible to Rhodes' and the Willow Springs, and fully expect to have the water there by the 1st day of June. This work will open one of richest and most extensive mining regions in California.

Negro Hill Ditch

The water ditch on the north side of the South Fork was smaller than its counterpart, the Natomas Ditch, on the opposite side of the river. As far as I can determine, it was completed later than the Natomas Ditch and may have been referred to as the Salmon Falls Ditch for a time. The ditch primarily served the mining community of Negro Hill. Slightly past Negro Hill, following the contour of the land, the ditch turned north and may have reached all the way to Condemned Bar. There are 19th century references to pipelines from the North Fork Ditch to supplement the water of the Negro Hill Ditch. While bridges to Condemned Bar and Massachusetts Flat could have supported a pipeline, I was unable to verify that such a conveyance system occurred.

Sacramento Daily Union, Volume 7, Number 1060, 16 August 1854

River Mining – Hill Diggings - Excitement on the North Fork — Granite Quarry. Mormon Island, August 10th. Messrs. Editors: This place takes its name from an island in the river opposite to the town, which was worked by Mormons in 1849, since which time it has been reworked every year, and is again this year paying wages.

The river companies are making extensive arrangements to work the river, and from what information I can obtain, with more than ordinary prospect of success. They certainly deserve to be rewarded as it is hard work.

Some of the hill diggings in this neighborhood are paying well, yet the general cry is hard times. A petition is in circulation to the Natoma Ditch Co. to reduce the price of water one-half, that is to say fifty cents per inch; setting forth that their claims are not paying wages after paying for water, and if their request is not granted many will have to abandon their claims.

There is some excitement here about new diggings on the South Fork of the American river, at and near Condemned Bar, three miles from this place. The diggings are very extensive, and will pay good wages. Two ditch companies are about carrying water there, one Jennings & Frazer, the other called the Salmon Falls Ditch Co. Messrs, Jennings & Frazer will have the water there in about two weeks.

There is quite an extensive granite quarry here at Mormon Island, situated in a ravine a short distance from the village, worked by Griffith & Richards. They furnished Adams & Co. of your city for their new building they cut it on the ground the exact size and shape. I am told they got it delivered in Sacramento very reasonable by teams going down light. It is said to be superior to the celebrated Quincy granite of Massachusetts, being free from all kinds of foreign substances.

Folsom Dam

Work on Folsom Dam began in 1948 and was completed in 1956. Dam construction of the 1950s obliterated much of the landscape around the confluence of the North and South forks of the American River. It is hard to distinguish what roads and ditches were a product of dam construction and which might have been on the dry lake bed before Folsom Dam. Even with as low as the lake level was in 2015, you had to hike several miles upstream before you encountered flowing water and could physically touch the river bed. On December 4, 2015, Folsom Reservoir reached its lowest elevation at 348.68 feet. At this level the lake was approximately at 14% of capacity. Just three months later, on March 8, 2016, after numerous rain storms, the reservoir elevation was 439 feet, the dam was releasing 8,000 cubic feet per second (cfs), and the inflow was 23,005 cfs from the North and South Forks of the American River.

2. Folsom Dam releasing 15,000 cubic feet per second on March 8, 2016, just a little over three months after hitting a historic low capacity.

Folsom Dam to Mooney Ridge

Beals Bar

Beals Bar was virtually at the confluence of the North and South Forks of the American Rivers. It is permanently under water at the foot of Folsom Dam. The closest one can get to the dam is above Beals Bar. When the lake level drops below 355 feet, you can see the piles of rock and drainage cuts as a result of the dam construction. It was in this approximate area to the west of the dam where the diversion tunnel for the American River was constructed. The American River was diverted out of its river bed and funneled into the diversion tunnel so the dam foundation could be constructed. What are visible to the hiker are the west wing dam and the concrete of the main dam rising out of and above the original American River channel. While it is always fun to gaze at a feat of engineering, the actual dam is as ugly as the destruction of the river from gold mining.

Weekly Alta California, Number 35, 30 August 1849

SUCCESSFUL GOLD DIGGING. — Dr H. Van Dyke, a member of the North Fork Dam and Mining Association, which company has recently completed a lateral canal at Beal's Bar, a little above the juncture of the North Fork with the Rio Americano, has just returned from their scene of operations. The work of drainage had been completed only three days before he left, and though the company labored under many disadvantages, they had raised in this short time over $15,000. This association is composed of about thirty hard working men, and from the result of the few days' labor since drainage, and the fine prospects of continued success, they confidently count upon a yield of about ten ounces per diem, each man, during the next and succeeding month. In confirmation of these statements we are at liberty to refer the reader to Mr. R. Van Dyke of the house of Bleecker, Van Dyke and Belden, of this place, where specimens of the gold thus obtained may be seen.

3. Branch water ditch from the main North Fork Ditch that serviced Beals Bar reservoir. Picture taken looking southeast toward the West Wing of Folsom Dam.

After gold mining faded away, the communities that did survive transitioned into more agriculturally based operations. Directly to the east of Beals Point, there were a couple of farms, and the remnants of foundations and driveways are still visible when the lake drops below 375 feet in elevation.

4. Distribution pipe from the main North Fork Ditch to a property owner west of Beals Bar.

Rattlesnake Point shows up on the 1954 Folsom USGS Topo map as fairly steep river bank east of Beals Point on the west side of the North Fork of the American River.

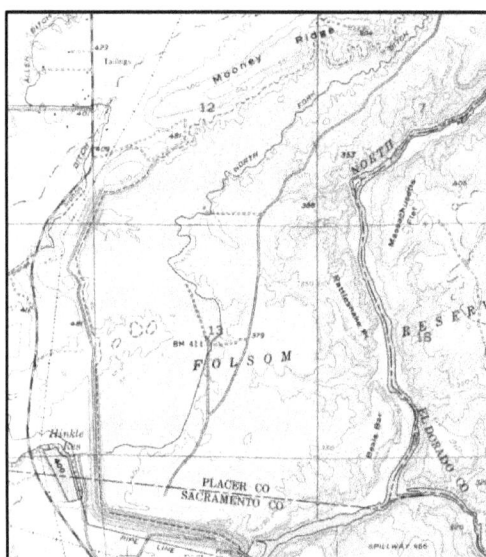

1954 USGS topographic map indicating Beals Bar and Rattlesnake Point. The North Fork Ditch is being directed into Hinkle's Reservoir as the North Fork Ditch Reservoir was partially covered by the West Wing of Folsom Dam.

I sloshed through some very shallow water at the lake's edge in November, but I never saw the steep drop off indicated on maps known as

Rattlesnake Point. If you look at the elevation lines on the 1954 map, the slope from the top of Rattlesnake Point was almost as steep as the sides of Folsom's dams and dykes.

North Fork Ditch Reservoir

Even though dam construction has heavily altered the landscape around the dams and dykes of Folsom Lake, there are still plenty of 19th century relics to visit when the lake is low. Before Folsom Lake there was the much smaller North Fork Reservoir that received water from the North Fork Ditch. It served a similar purpose as Folsom Lake in that it stored water from the American River for delivery later. North Fork Ditch reservoir lies just to the west of the confluence of the North and South Fork on a plateau above the river at approximately 400 feet in elevation. Initially completed in 1853, the North Fork Ditch wound its way from Auburn along the North Fork of the American River, terminating west of what was Beals Bar.

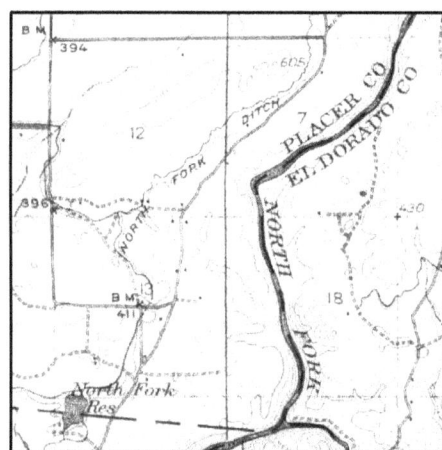

1941 USGS topographic map shows the North Fork Reservoir straddling the Placer and Sacramento County boundary west of the confluence of the North and South Forks of the American River.

The North Fork Ditch reservoir is illustrated on several maps. The west Folsom Wing Dam wrapped around the old reservoir. The old North Fork Reservoir was completely filled with rock and is only distinguishable as a football field-sized plot of granite rubble at the base of the west wing dam lakeside.

5. Looking west, this North Fork Ditch concrete structure diverted water toward Baldwin and Hinkle Reservoirs. The West Wing of Folsom Dam is at the top, and the filled North Fork Reservoir shows below the dam in the upper left-hand corner.

However, there were many parts of the North Fork Ditch that were not destroyed by the dam construction. Still standing are concrete canals that would direct water from the North Fork Ditch either into the reservoir, east to the Beals Bar branch canal, or westward to Baldwin Dam reservoir. Baldwin Reservoir was the main source of water for the development of Orangevale in the late 19th century. Hinkle Reservoir, still in use today, was built to replace the North Fork Ditch Reservoir. The reservoir now holds the treated lake water for San Juan Water District before it is pumped to the

communities of Folsom, Orangevale, Granite Bay, Fair Oaks and Citrus Heights.

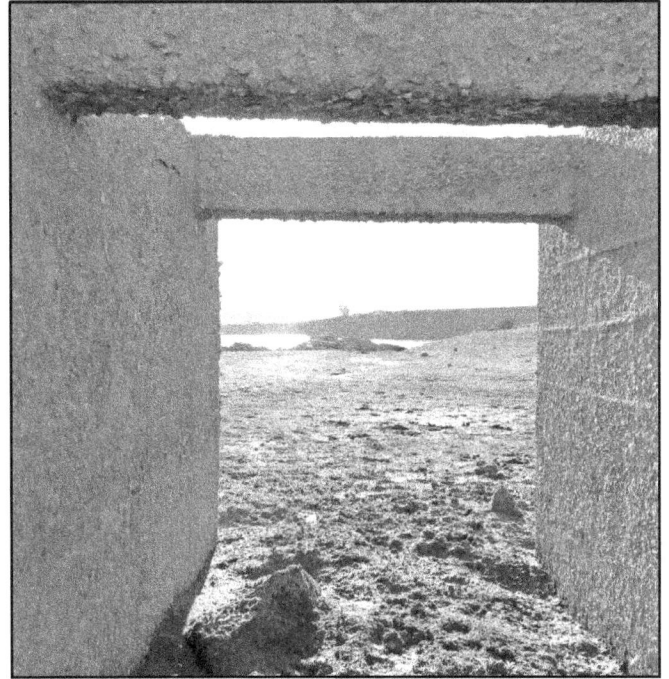

6. Looking east at Folsom Dam. Through this concrete structure, water from the North Fork Ditch was allowed to flow toward the Beals Bar reservoir.

You'll notice on the 1954 Folsom USGS topographical quadrangle map that the North Fork Reservoir is no longer depicted as it is in the 1941 Folsom USGS map. But the North Fork Ditch represented, by a blue line right above the 400 foot elevation line, is shown going under the wing dam around Hinkle Reservoir over to Baldwin Reservoir. From the 1952 aerial photograph of the Folsom Dam construction it looks like the west wing dam was left open at this spot where the North Fork Ditch was delivering water to Baldwin Reservoir under the wing dam's base. This arrangement of a tunnel under the west wing dam was probably terminated once the pipelines from the face of Folsom Dam on the downstream side were completed and connected to the San Juan Water District treatment facilities.

7. Looking southwest toward the West Wing Dam. This nearly soil-filled concrete canal directed water into the North Fork Reservoir.

If you look closely on the lake side at the base of the west wing dam, you can make out where the old North Fork Ditch use to travel under the west wing dam over to Baldwin Reservoir. Baldwin reservoir impounded water for delivery south to gold mining operations on the lower American River, west to Orangevale Colony, and north in the Rose Spring Ditch for mining and agricultural operations.

Sacramento, Placer, and Nevada Railroad

Visible on a 1952 aerial photograph is a road that runs in between the American River and the North Fork Ditch. From research that I and others have done, there is strong evidence to suggest this was the original railroad grade of the Sacramento, Placer, and Nevada Railroad (SPNRR). A very short-lived rail line, the SPNRR ran from Ashford on the north side of the American River (Greenback Lane and

Auburn-Folsom Boulevard) up to approximately a mile north of King Road in Placer County. It operated from 1861 to 1864.

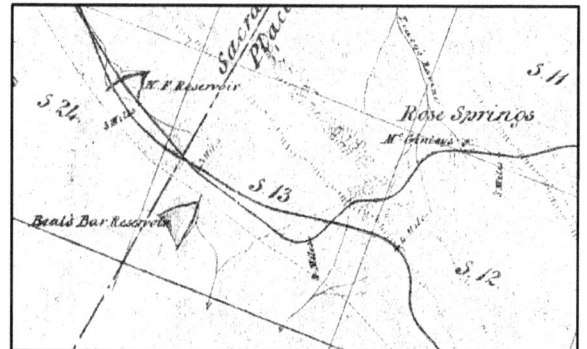

This section of the 1861 map of the Sacramento, Placer & Nevada Railroad depicted two alternate routes. The final grade turned northwest by Rose Springs. Note that both the North Fork Reservoir and the Beals Bar Reservoir are shown on the map.

The SPNRR was referenced on a 1913 map of Placer County drawn by L. F. Warner Jr. and which hangs in the Placer County Archives office. Even though this map was printed, not on paper but on a heavy canvass material, and was well worn, the dash line and associated title *Sacramento And ....R.R.* is still visible. I'm not sure why the cartographer would include a rail line that had not been in existence for almost fifty years on the map. But it shows the line entering Placer County at Section 13, Township 10 North, Range 7 East. The outline of the rail line and officially marked roads mirror the 1941 USGS Folsom topographical map for roads and trails in that region.

Daily Alta California, Volume 13, Number 4066, 4 April 1861

Railroads. — During last week, the Sacramento, Placer, and Nevada Railroad Company concluded a contract for the iron and laying the

track from Folsom to Auburn, and the Sacramento Valley Railroad Company have engaged to run their machinery upon the road. The subscription to this road nearly equals the estimates for grading; the first division of thirteen miles. The Company have concluded to pay in cash for grading, engineering, etc., and will issue bonds in payment for the iron only; but to enable them to carry out their original idea, some $7,000 or $8,000 more should be subscribed.— Sac. Bee

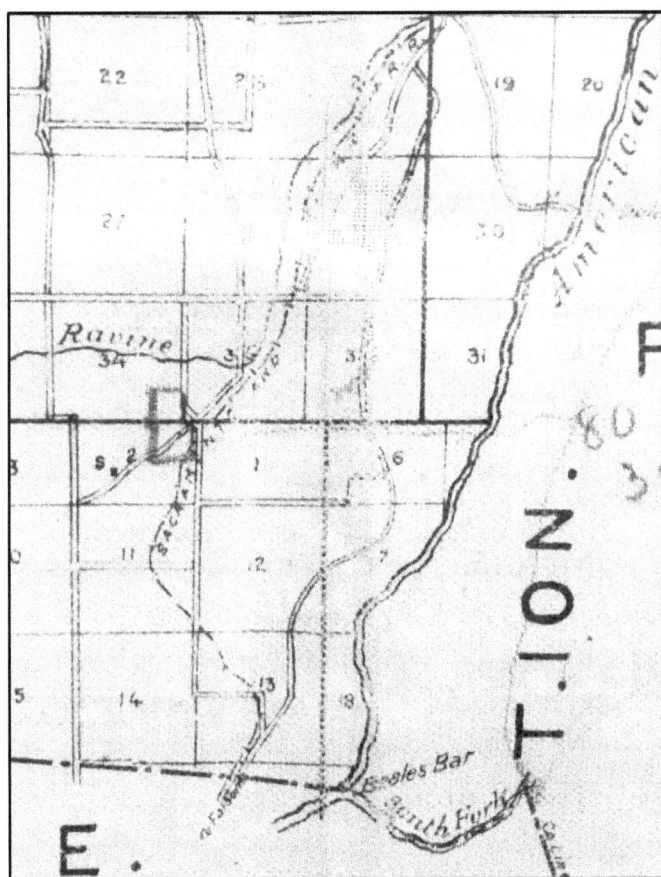

1913 Placer County map showing the railroad grade of the SPNRR.

The SPNRR would have entered the Folsom Lake area traveling north at the west wing dam and then traveled around what today is Beals Point State Park. It then followed a branch canal of the North Fork Ditch called the Allen Ditch

that delivered water northwest up to Miner's Ravine Creek. (On none of the maps I have reviewed was there a specific name given to the western branch of the North Fork Ditch at this point. There are references to the Rose Spring Ditch that travelled south and George Ditch that traveled north. The 1954 Folsom USGS topographical map names the remaining water ditch, post-dam construction, as the Allen Ditch. I have chosen to use that title for reference purposes in this book.)

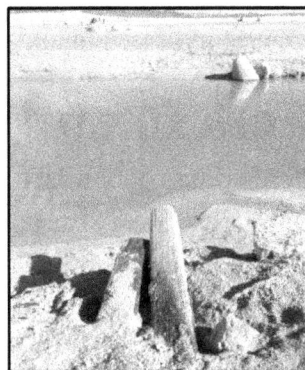

8. Bridge timbers poke out of the earth south of the Dyke 5 drainage cut. This would have been the approximate location where the SPNRR met with the Allen Ditch. There was a dirt road that crossed the Allen ditch at this location shown on a 1941 USGS topographical map.

Both the Allen ditch and the railroad travelled east between Beals Point to the south and Gold Mine Ridge to the north. What is important to remember is that the water ditch needed to maintain a very gentle, nearly flat slope, because it was gravity that propelled the water in the ditch. Railroads also like gentle slopes, so it was natural for the railroad to follow a grade that had already been established by the water ditch.

After the railroad rails were removed, the grade eventually turned into a paved road. The sections of the old paving can still be seen when the lake level drops below 400 feet in elevation. The

1941 USGS topographical map indicates that north of Boundary Monument (BM) 411, the road becomes dirt as it travels to meet up with the Allen Ditch and skirts to the south of Rose Springs and north up to Miners Ravine. If you look close at the 1941 map, you will notice a road is marked between two dirt roads on either side of the Allen Ditch east of elevation marker 396. I believe this was a bridge crossing the ditch and connecting the two roads. I photographed large square-cut timbers that could have been remnants of the old bridge near the location where a bridge was indicated on old maps.

North Fork Ditch 1915

In the early 1910s it was determined that the North Fork Ditch was losing a substantial amount of water to leakage through the earth-lined canal. A program was put in place to line the entire ditch with concrete. It is the concrete ditch lining that can be seen today when the lake is low.

9. Looking north toward Mooney's Ridge, the concrete lining of the North Fork Ditch is visible.

Sacramento Union, Number 53, 23 October 1915

AUBURN (Placer Co.). Oct. 22.—A large force are in engaged in lining with cement the ditch of the North Fork Water company, in the canyon of the American below this city. The cement will prevent a large amount of seepage, thus giving the irrigators more water during the season.

10. The year 1915 was ceremoniously etched into the top concrete distribution box on a branch of the North Fork Ditch.

It is on a branch of the main North Fork Ditch that I found a rare date impressed in the concrete. At what might have been a distribution box to meter water to a farm or mine, someone had etched 1915 onto the top of the concrete lining. As far as I can ascertain, the 1915 etching was on a branch line of the main North Fork Ditch. There were numerous branch canals that served mines and farms off of the main water ditch.

11. Half-moon concrete support for a metal flume of the North Fork Ditch north of Beals Point.

Around Mooney Ridge and north of Beals Point the North Fork Ditch runs at an elevation of 408 feet. This put the water canal up to a mile away from the actual river and 50 feet higher than the farms closer to the river that were at about 350 feet in elevation.

The North Fork Ditch was engineered to allow the water to move down the ditch at a predictable rate through the force of gravity. Consequently, a specific grade or slope had to be maintained over the course of the ditch. Because it is much less expensive to cut a ditch into earth than build wooden or metal flumes, the water canal closely follows the natural contours of the landscape wherever possible. The North Fork Ditch started out at an elevation of approximately 550 feet. The diversion dams were next to the City of Auburn and terminated in the North Fork Ditch Reservoir at approximately 400 feet elevation. There were places where flumes were employed to span short sections across creeks or deep ravines.

Sacramento Daily Union, Volume 9, Number 1262, 10 April 1855

North Fork Canal. This canal is located on the north side of the North Fork of the American river, in Placer county. Its length is a little over thirty miles, and it waters some of the richest and most extensive mining districts lying on that famous river. The dam which turns the water into the canal is built at Tahmaroo Bar. The first point passed is Oregon Bar, then follow in succession, Rattlesnake Bar, Horse Shoe Bar, Doten's Bar, Condemned Bar, Beal's Bar, Slate Bar, to Mississippi Bar in this county. At all these Bars rich diggings have been found in the bluffs, or bars as they really are, although from fifty to one hundred feet above the river bed. They are composed of cobble stones, boulders, clay, sand, and cement, lying on granite, where they must have been deposited by the action of water thousands of years ago. These deposits are generally found in a kind of basin, of various depths, and not unfrequently pay very rich. Since water has been carried on the bluff above these bars, they are staking them down in many places from ten to thirty feet in depth, the dirt often paying from the top down. Some claims, however, only pay for a few feet near the bed rock. Between the points we have named, there are numerous hills which are paying finely. There are in all some forty distinct mining districts watered by this canal from Tahmaroo to Mississippi Bar. The ground is not yet half prospected, but from what has been done, it is evident that it will take years to exhaust them.

At a point just above Beal's Bar the canal reaches the divide, and the Company can run their water over towards Bear River[?], and supply the Rose Springs diggings, at which point, it is confidently stated, there are at least ten acres of deep and rich mining ground.

This canal is carried over one of the roughest lines in the country. For some two years it has been considered next to impossible to take the water out successfully at that point, but the present Company, with Mr. Eliason as engineer, and Messrs. Brooks & Clark as contractors, soon demonstrated its practicability. There is comparatively but little fluming on the line, the contractors preferring to make a canal where it could be done, as a canal once well built is good for all time ; whereas, a flume will only last about three years. In many places the lower side of the canal embankment is supported by a wall of granite rock.

Probably no work in California of like magnitude has ever been completed in the same time. It is eight feet wide on the top, four on the bottom and three feet deep, with a fall of four feet for the first half and five for the other, and carries water enough to supply two hundred and fifty sluice-heads. The capital stock has been lately increased and is now $300,000. The stock promises to become valuable. It was commenced in September last, and the water was running nearly its whole length in about six months from that date.

The Company have nearly completed ten reservoirs, which hold all the water which the canal can carry during the night. A. P. Catlin, Esq., the President of the Natoma Canal Company, is also the President of the Forth Fork Canal Company.

Around Beals Point wave action from the lake has undermined and eroded the soil along this stretch of the North Fork Ditch. The concrete lining has flopped over on its side. In other places where the ditch had to be dug through the soil and below the landscape surface, only the top of the concrete lining may be visible.

Further up the North Fork of the American River, where the North Fork Ditch hugged hillsides or granite outcroppings, the builders would reinforce the canal with a rock retaining wall.

Then there were times when it was just easier, albeit more expensive, to flume the water across a creek or ravine. At BM 411 on the 1941 USGS Folsom topographical map, you'll see that the North Fork Ditch had to cross a little ravine. The brown line is the 400-foot elevation contour line. When the water is low, you can see the half-moon concrete support. This would have been for a half-pipe wooden or metal flume to carry the water across the little gully and back into the concrete-lined canal.

Dyke 5 Drainage Cut

North from the intersection of the old county road and the flume crossing, northeast of Beals Point, is a small hill. To the west will be Dyke 5 of Folsom Dam; to the north is Gold Mine Ridge that connects to Mooney Ridge via Dyke 4. Gold Mine Ridge and Mooney Ridge are higher in elevation than dykes 4 or 5, and constitute part of Folsom Dam. To the east are the lake and the river channel of the North Fork. At the bottom of the swale is a man-made drainage cut that starts several hundred yards from the base of Dyke 5 and travels toward the lake at low water elevation. The cut is visible when the lake drops below 380 feet in elevation.

12. From on top of Dyke 5, looking east, the drainage cut from the base of the dyke can be seen directing water to drain toward a shrunken Folsom Lake.

The topographic maps reveal a slight rise of the topography between the river to the east and the base of Dyke 5. Without this drainage cut, as the lake recedes during its normal drawdown during the summer, a small lake would have been formed. The cut, I assume, was constructed to drain the water away and prevent fish from being trapped. This cut also bisected the North Fork Ditch and the unpaved county road.

Dyke 5 also prevented the Allen Ditch, which was a spur from the North Fork Ditch servicing farms to the north up to Miner's Ravine Creek, from continuing to operate. The Bureau of Reclamation improved the existing Rose Spring Ditch west of Dyke 5 with siphons to continue water service. (Rose Spring Ditch started near Baldwin Reservoir and travelled north around present-day Folsom Lake Estates and the City of Roseville Water Treatment Plant on Barton Road.)

13. The drainage cut is approximately 15 to 18 feet deep in some places.

Rose Springs

Months into the drought of 2015 I still had to hop over the drainage cut that always seemed to be inexplicably filled with a little ribbon of water heading toward the lake. At this point in the drought, the lake shoreline was over half a mile away to the east, and the lake bed had been baked dry by the sun for months. I traced this little creek of water to the base of Gold Mine Ridge directly east of Dyke 5. Here were two little pits, full of water, spilling over toward the pond at the base of Dyke 5. The pond was ringed with grasses, along with Canada geese and killdeer, who had found an oasis to call home in the summer heat and drought.

14. Looking north towards Beals Point, one of the Rose Springs bubbles water down the side of the hill to the base of Dyke 5.

Several old maps indicate Rose Springs was a collection of buildings west of the North Fork of the American River. There are also accounts of an active mining district, once the North Fork Ditch provided water to wash the dry diggings in the area.

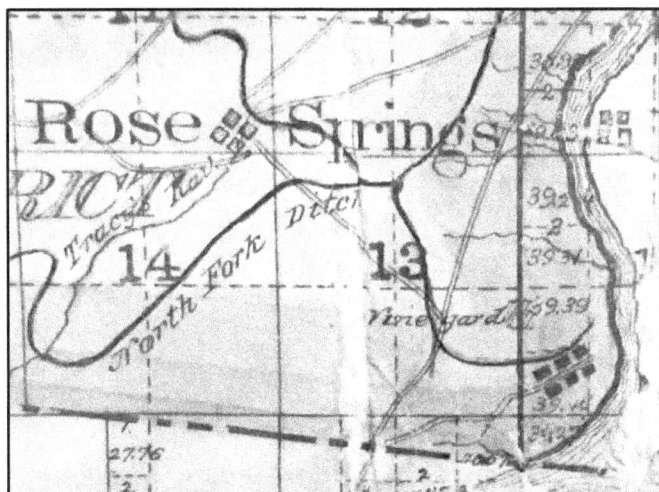

1887 Placer County map showing a small community that had been established around Rose Springs.

My novice map-reading skills always drew me to the conclusion that Rose Springs was west of Folsom Lake's Dyke 5. It wasn't until I reviewed a 1952 aerial photograph from before the dykes were finished that I noticed a trail of dark vegetation in the black-and-white photo that led me to believe otherwise. In addition, I found a dump site where the small Rose Springs community is shown on an old map. The 1887 Placer County map and 1910 Water Canal map indicate several buildings at this site. Also indicated on several maps is a road that passed through Rose Springs and then closely paralleled the Allen Ditch northward. The 1941 Folsom USGS topographical map does indicate one building east of BM 396 along the county road.

The maps and physical evidence have led me to conclude that the bubbling hole of water on the south face of Gold Mine Ridge, east of Dyke 5, is one of the original Rose Springs.

*Allen Ditch most likely served the Allen District Mines, which, according to "The Historic Sacramento to Auburn Road, from Miners Trail to Interstate Highway," by Leonard

M. Davis, 1996, was a mining area six miles east of Roseville. This may be the site of the quarry ponds to the south of Douglas Boulevard today.

15. At the height of the 2015 drought, Rose Springs continued to produce enough water to fill a small pond and create wetland habitat for many birds at the base of Dyke 5.

Before the construction of the North Fork Ditch and easily available water along the water canal, naturally occurring springs were an important source of water for miners and livestock. Even though the river was only one mile away, there was still a steep slope to walk in order to reach water's edge. A small bubbling spring could provide a little water and shade from surrounding trees. The near year-round production of these springs was important enough to note on maps for travelers and as reference points on the map.

From all this evidence I have concluded that the little puddle of water that continued to flow, even during one of the worst droughts of modern California history, was indeed Rose Springs. The Rose Springs outlet sits at approximately 415 feet in elevation and must be supplied by water draining from the Gold Mine Ridge above it that reaches 525 feet in elevation and is approximately a quarter-mile long. Rock Springs is on the eastern toe of Mooney Ridge and also continued to bubble water late into the summer of 2015.

Rose Springs Dump Site

An archeologist's dream is to stumble upon a trash pit or dump site in order to search for clues about previous regional inhabitants. The Rose Springs dump site does not have a sign unless you consider the scattered shards of glass around the area to be an alert. It is not unusual to find broken glass on a dry Folsom Lake bed. But resting on the surface in this area was a variety of different types of colored glass, pottery, and china. I started digging around and found the general location and outlines of what might have been a community dump site.

I placed a variety of broken colored glass and ceramics on a granite rock nearby the Rose Springs dump site.

I placed larger pieces of the glass, bits of old plates, metal from what looked like an old tractor on some granite rocks just northeast of the dump site, within walking distance. When I returned to the dump site in 2015, most of the glass and metal had disappeared from where I had placed it.

Most likely a movement from a Big Ben alarm clock found at the Rose Springs dump site. I have several in my clock collection that are in much better, and working, condition.

The one item I could identify, thanks to my previous hobby of collecting old mechanical clocks, was the rusty movement for a Big Ben alarm clock. This type of alarm clock movement was made anytime from the early 1910s up through the Second World War.

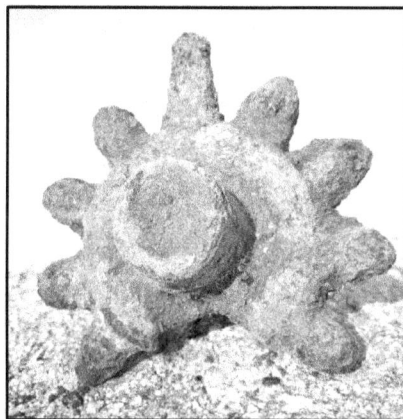

A rusty gear, approximately 6 inches in diameter, from some sort of a machine found at the Rose Springs dump site.

Granite Bay Gold Mine Ridge

A layer of river cobble and mud sits atop a layer of granite north of Beals Point. Miners would dig down through the river cobble to the granite base hoping to find gold.

From Beals Point you can see how the volcanic debris and mud flows sit on top of a granite pluton on Gold Mine Ridge directly to the north. Erosion from Folsom Lake has stripped away loose soil on the ridge's south side revealing solid granitic rock. Twenty to thirty feet of unconsolidated smooth river cobble and mud sits on top of the granite ridge. On top of Gold Mine Ridge was the opening of a gold mine whose exit was on the north side.

16. The pit to the gold mine was covered by some chain link fence to keep animals and humans from tumbling into it.

From on top of the ridge, in alignment with the main pit on top and the exit on the side of the hill, are a series of depressions and mounds of rocks. These might have been other shafts or test holes as the miners work their way across the granite base.

The Bureau of Reclamation pumped thousands of gallons of water into the open pit of the gold mine during the summer of 2013 in an attempt to find out where the water might surface. In the course of strengthening the dykes and dams around Folsom Lake, the Bureau had stumbled across several old mines near the dykes through this unconsolidated material.

17. Before being filled in by the Bureau of Reclamation, the gold mine pit was approximately 10 feet deep.

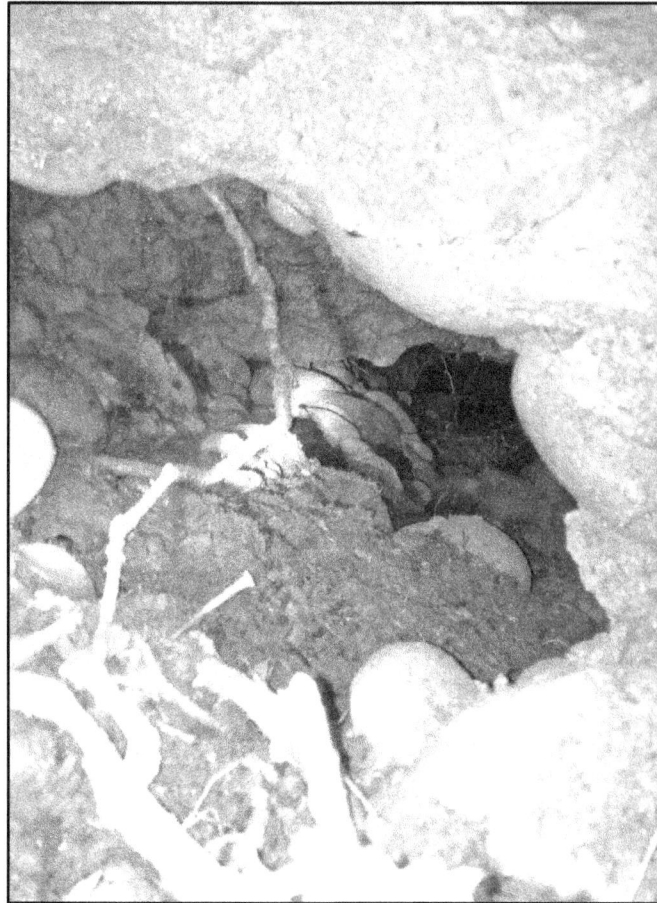

Exit, or maybe entrance, to the gold mining shaft.

They wanted to find mine shafts or cavities that might potentially undermine the dykes. I was there when they pumped the water into the pit on top of the ridge. No water was ever seen visibly leaking from the sides of the ridge or dykes. Of course, Rose Springs, at the southern base of the ridge, continuously leaks water, possibly from the entire ridge.

Base of Mooney Ridge

The filling of Folsom Lake and subsequent wave action of the water has helped erode a tremendous amount of soil from the high-water marks of the lake. It is not uncommon to come across parts of the North Fork Ditch completely filled with decomposed granite.

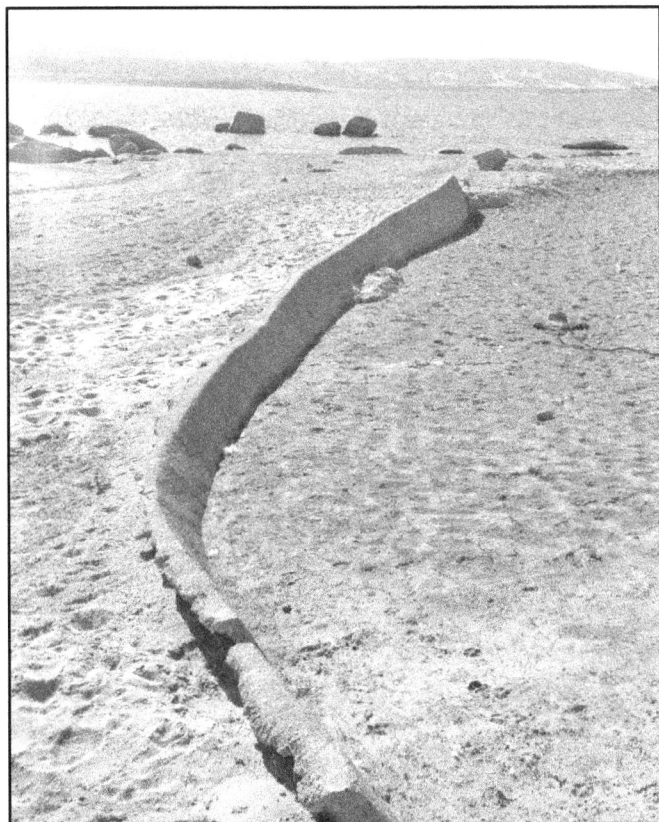

18. The remnants of the concrete lining of the North Fork Ditch curves westward around a hill to keeps its elevation grade. Most of the canal has been filled with decomposed granite from the hill above, and only a lip of concrete is visible.

The only way you can identify that the water canal existed at all is if you see the top of the concrete lining poking through the surface. There is one part of the ditch that is exposed, and at its base you can see the square hole used to release water down the hill to a farm or mining operation.

19. Water from the North Fork Ditch flowed through a square hole in the concrete lining. At this service connection the water was directed through a wooden sluice into a metal stand pipe.

One of these water service connections sent water down the hill to an orchard on the bluff of Rattlesnake Point. When the lake level gets very low, you can see the stumps of the trees along with the foundations of barns or other out buildings. If the lake level drops below 370 feet elevation, you'll be able to make out the old county road at the base of the old orchard.

20. The concrete outlines of a building next to an orchard, as indicated by the rows of tree stumps that received water from the North Fork Ditch.

21. Parts of the North Fork Ditch were supported with local granite rocks.

22. In order to maintain the gently sloping grade, the North Fork Ditch was cut into granite outcroppings shown here at the eastern tip of Mooney Ridge.

Native American Grinding Holes

One of the most exciting discoveries I made came when the lake was at its very lowest in November 2015. While walking around some granite boulders along the shoreline at the southern tip of Mooney Ridge, when the lake level was at approximately 350 feet of elevation, I stumbled across some Native American grinding holes. What was unique about these grinding holes was that they weren't on top of a large flat granite boulder, which is where I normally find them. These holes were on top of a small boulder about waist-high, almost like a kitchen countertop. Plus, there was another boulder leaning on the grinding hole work station that would have provided some shade from the morning sun.

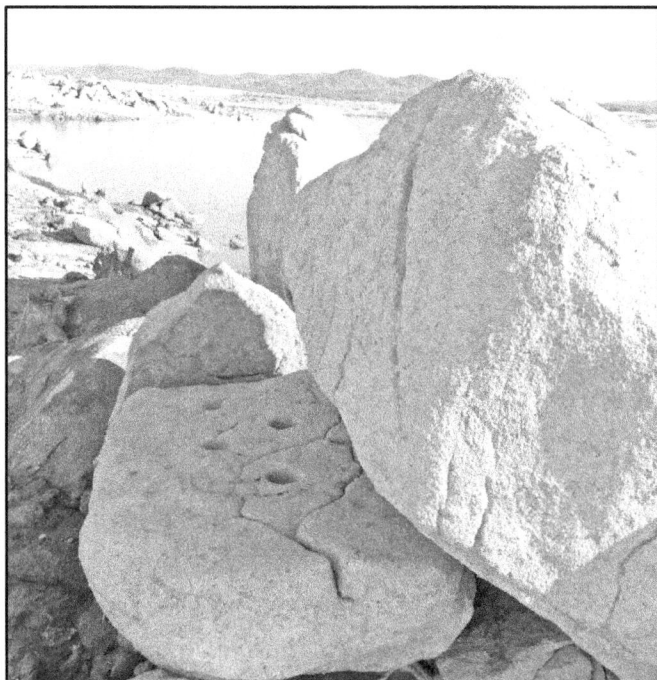

23. Native American grinding holes found off the eastern tip of Mooney Ridge.

Normally the grinding holes are situated on top of large granite boulders, where the large surface area would allow the worker to sit and smash the acorns in the hole. These grinding holes seem to indicate they had been used in a standing position, as there is not enough room to really sit comfortably. Regardless of how they were used, it was a pretty cool reminder that European immigrants were not the first to use the resources of the river. Further south of the Native American grinding holes, the piles of river cobble tailings from mining operations became visible.

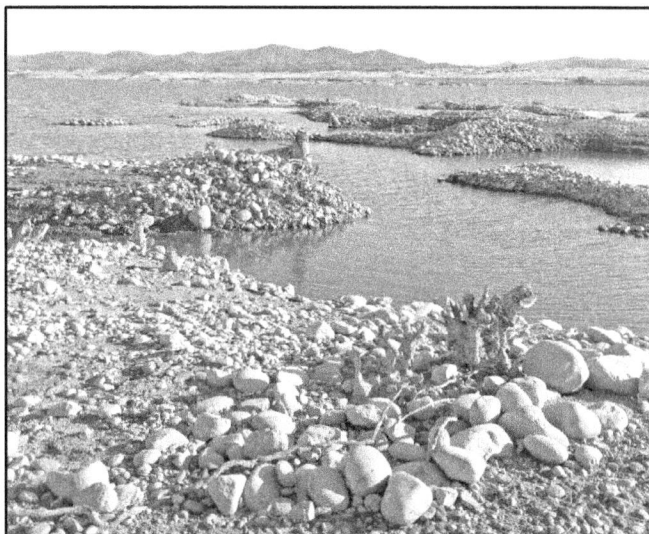

24. The tailings from gold mining operations south of Mooney Ridge stretch off into the distance.

Rock Springs

North of the grinding holes at the east end of Mooney Ridge are the foundation and front porch pillars of an old home. The 1941 Folsom USGS topographical map indicates this house was at approximately 375 feet elevation. The 1910 water canal map lists this building as Rock Spring. While the house or building was next to the county road, the actual spring is north and up the hill approximately 200 yards at about 380 feet of elevation.

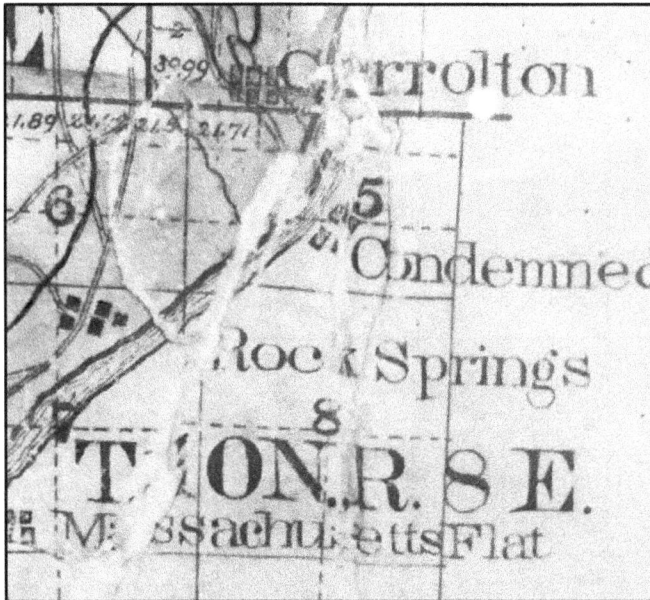

Rock Springs listed on an 1887 Placer County Map.

Rock Springs was bubbling out water late into the summer of 2015. It created a green trail of grass as its little output of water meandered toward the shrinking shoreline of Folsom Lake. It is located near a favorite parking spot for cars and trucks when the lake is low and frequently gets driven over. If you look closely when Rock Spring is bubbling up water, you'll see grains of sand being tossed about in the shallow puddle of clear water.

25. Similar to Rose Springs, but with less volume, Rock Springs continued to bubble up water late into the summer of 2015.

There aren't many references to Rock Spring, but I did happen to come across a description of the spring accompanied with the cathartic confession of a miner's death.

Sacramento Daily Union, Volume 3, Number 223, 3 November 1877

A MYSTERIOUS LETTER

The following letter, postmarked "Philadelphia (Perm.), October 18th," will be read with curious interest by old residents of the places named therein. Captain William Siddons, one of the parties whose names is used, identifies the other parties named, and also has an imperfect remembrance of the Dr. Lane, supposed to have been murdered.

To the Sacramento Union: As your paper was as familiar as household words to all early Californians, and for many years I have cut loose from all communication with your State, fraught with its terrible recollections, I am impelled by a fatality I can not more definitely explain, to bring to light, in this manner, an episode which has been the blight of my life, and which has been locked up securely in my breast alone, for a period of sixteen years. The imbruing of one's hand in the blood of a fellow-miner was not so common an occurrence as picture life of early days in California is generally depicted, and was looked upon with the same horror, and guilty terrors followed its perpetrator with more pertinacity than elsewhere, where the population is not so migratory, and the chances fewer of meeting with acquaintances made in former camps.

In 1852 I arrived in Hangtown, after crossing the plains, and my first mining was done at Redbank, on the South Fork of the American river. Afterwards I took up a claim on Swindling Hill, near by, and with a partner, a very tall young man, named Mose, and in the first month washed out a hidden can containing $5,000. From thence I went to Rattlesnake Bar, on the North Fork of the American river, boarded with a Dr. Frey for several months at a miners' hotel, and prospected up the river, making many acquaintances, among whom was a Theodore Beecher, claiming to be a cousin of the famous Henry Ward Beecher, with whom I became intimate; also Tom Merrill, a billiard sharp ; William Siddons, a bar keeper, and a family at an isolated spot some few miles off called Rock Springs, whose waters, of coolest temperature, was the resort of swarms of humming-birds, which fluttered among the surrounding foliage incessantly as I there was -regardless of the approach of mankind. I mention these and similar names and description of places, as they may be recognized by early residents, and will verify my concluding statements. Tamarou, Manhattan, Murderers' and other bars consecutively engaged my time until '61, I found

myself at Michigan Bluffs, still further up the river, and where it was my fate to become acquainted with Cap. Lane, a tall, elderly man, with very long, dark whiskers, which covered his whole face; he was a misanthropic person—living always alone in his cabin, only issuing from it on rainy days for his frugal allowance of salt pork and Bayou beans. I had been prospecting near his cabin in an unfrequented ravine, when he strolled along with his big gum boots and dollar suit of overalls and jumper, and I stopped him with a customary salutation. We had chatted but a minute when a gust of wind blew his whiskers to one side, uncovering a part of his face comparatively bare, when thought I saw a letter branded there. Involuntarily, and without intent to insult him, I mentioned the discovery, when his rage became so unbounded that he threw himself upon me with no weapon but his talon-like fingers, and we both fell from the bluff bank and landed at the bottom— he underneath. I disengaged myself quickly, and was about to strike, when I found he did not get up or attempt to move, and after waiting a reasonable time for the renewal of hostilities I drew nearer, and on examination, discovered that he was indeed dead! Had I then had the wisdom to have gone to the Justice (Cunningham), and boldly told the story and been honorably exonerated all might have been well; but to my excited mind appearances seemed so much against me that I determined on a much more hazardous proceeding, and with a, superhuman strength I carried him the short distance to his cabin, laid him on his bunk as composedly as though asleep, and hastened to bid adieu to the camp, without knowing if his sudden death was caused by a broken neck or by the violence of his sudden passion. Neither have I had temerity to inquire since, through any channel, of the supposed cause of his death, and, though eagerly scanning for years the Union, the Forest Hill and other papers which chance placed in my way, never have I seen anything relating to the occurrence. At times there is a ray of hope inspires me, that in my excitement I had mistaken a state of coma for death, and that the old gentleman may yet be alive. And in such a spirit I indite this article, hoping against hope for a confirmation.

I will but hurriedly detail my course since that fatal day. I secluded myself for a week in the solitude of the deep woods of a big oak Hat on the opposite side of the river, having supplied myself with provisions from a Greek's store, and then by trails and cutoffs seldom used wandered through dense forests and crossed high mountains, to me unknown, but which can be understood by those acquainted with that country when stating that the first person to whom I ventured to ask my whereabouts informed it was Mosquito canyon. I kept on without method or care, having my belt better stocked than I the usual careless miner, but which would have been freely exchanged for the light heart I bore when lying in the shades of Rock Springs, admiring the plumage of the humming birds. Silver creek was crossed at a perilous place, whose banks were 600 feet high, and where in descending its steep declivity I narrowly escaped with life, through treading on it mossy rocks, made treacherous by moisture, at one time sliding over 100 feet, wildly clawing at the slimy, glassy surface, until my fingers fortunately caught in a crevice. With a tortured mind and aimless existence, I made my way to Nevada and subsequently to Salt Lake City, and in turn to the mines of Idaho, avoiding Californians as much as possible, and at length worked my way East, where years of application to business have not weaned from recollections

of my fatal rencontre. I have carried my load so far alone, but hope this revelation may have the effect of bringing a quiet hitherto unknown. Coso.

Later in 1860, a correspondent for the Daily Alta California on a trip to see the Alabaster Cave in El Dorado County made this observation of Rock Springs.

Daily Alta California, Volume 12, Number 133, 13 May 1860

Five miles above Folsom is the Rock Spring house, on the site of which, eleven years before, a mining company consisting of twenty men, encamped amidst a mass of rocks and wood, prior to a descent on the rich bars within a mile of the spring. What a change has come over the scene since that day! We found here a comfortable looking house, and one of the thriftiest peach orchards in this section of the State. The grounds are splendidly watered, and the enterprising proprietor furnishes his table with vegetables from his own garden.

26. The pillars of a porch and concrete foundation are all that remain to be seen of the Rock Spring House when Folsom Lake drops.

Irrigation for orchards was no doubt supplied by the North Fork Ditch that ran slightly above Rock Springs and the Rock Spring House.

Granite Bay to the Narrows

Carrolton

Between Granite Beach State Park and Dotons Point there is very little in the way of historical sites to be readily found. The area has been heavily worked with large earth-moving equipment to construct the various boat ramps for the park. Consequently, any remnants of the North Fork Ditch have been destroyed. Several maps indicate a community called Carrolton existed at the base of today's Dotons Point. All I ever came across were piles of river cobble mine tailings and mining pits. It's possible that any building foundations still lay closer to the river, which was underwater even at the low level of 350 feet of elevation.

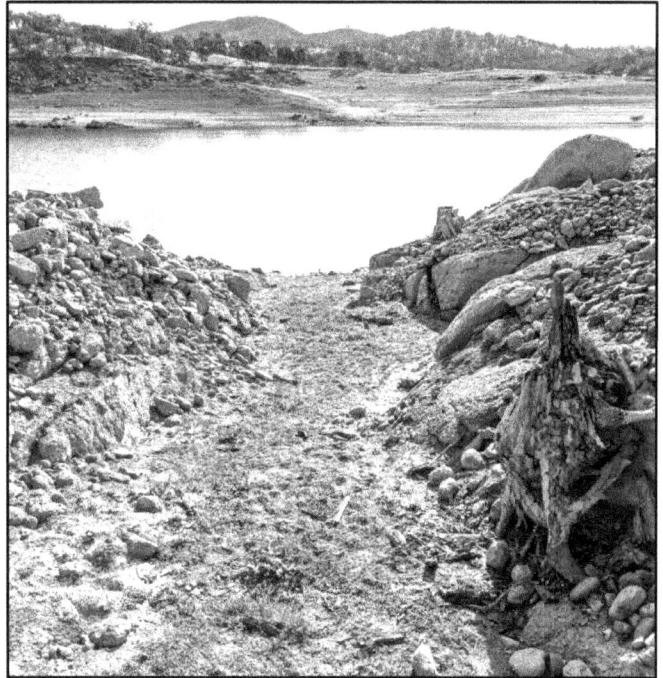

27. A clearing between the mine tailings near the old town of Carrolton looking east toward Condemned Bar on the Peninsula side of Folsom Lake.

The 1910 Water Canal map shows the community of Carrolton below where Folsom State Park's Dotons Point is today. The community of Dotons Bar is further north just east of the subdivision of Los Lagos, where the North Fork Ditch had to meander a quarter of a mile west to maintain its elevation grade. I'm not sure why Folsom State Park chose to name the point above the historic community of Carrolton after a river bar a mile and half north, but they did. It would have been more interesting to name the point and campground Condemned Point after Condemned Bar right across the river.

28. A fairly large weir for the release of water from the North Fork Ditch down to mines or farms around the Carrolton area.

Daily Alta California, Volume 6, Number 138, 1 June 1855

American River Towns. Carrolton, May 27, 1855

Four months ago this place, which now numbers three or four hundred people, contained but one house. It had been supposed that the richest places on the American River were exhausted: but new towns crow into life as if by magic on the route of permanent water ditches. This place has had water two months, and six stores, five boarding houses, two black smiths shops, two butchers, one carpenter's shop, two ten-pin alleys, two billiard saloons. Wells, Fargo At Co.'s express office do the exchange of the people. The miners have not yet fairly opened their claims, but are confident they will pay well.

Carrolton is about the centre of the North Fork Ditch, and is expected by the citizens to become the centre of business. Two stages from up the river on either side pass through both ways daily, and another that runs to Auburn, is to continue from Doten's and Carrolton, three miles to Slate Bar. The growth of this latter place is quite equal to the former. Slate Bar was started six weeks ago, and a friend, a miner, informs me there have been four dwelling houses for

families, and one store erected the past week, that the miners are congregating rapidly, the water having been in eight days, and the work has yielded from six to sixteen dollars a day.

Doten's is two miles above Carrolton, an older place by two or three months, a place where much talent has seemed to congregate. It is now the largest place on the river. Condemned Bar, opposite the river from Carrolton, is several months older and about the same size.

Beeks Bight

North of Dotons Point the river canyon starts to narrow. At this point the North Fork Ditch is at approximately 420 feet elevation. There are remnants of the North Fork Ditch traveling up the river canyon, but they can be hard to spot. The North Fork Ditch had to make a long detour north and west to maintain its nearly level descent through what is now the Beeks Bight area of Folsom State Park.

29. Foundation of an old building right along the elevation and path of the North Fork Ditch is due east of the Beeks Bight parking lot.

The trail from the parking lot at Beeks Bight is essentially the soil-filled North Fork Ditch. Directly east of the little point that is the entrance to Beeks Bight is the concrete foundation with steps. A building at this particular location is represented on the 1954 Auburn topographical map. It's interesting that the interior foundation of the concrete structure is formed with local rocks.

Dotons Bar

When the lake level is below approximately 420 feet in elevation, the North Fork Ditch makes an excellent hiking trail.

30. Granite boulders make the western wall of the North Fork Ditch above Dotons Bar.

You will be able to walk through parts of the ditch that were constructed with towering granite boulders as the west wall and poured concrete on the river side.

31. A large tree grows out of a weir outlet of the North Fork Ditch that supplied water to mining operations on Dotons Bar.

There are also outlets in the canal that allowed water to flow down to mining operations.

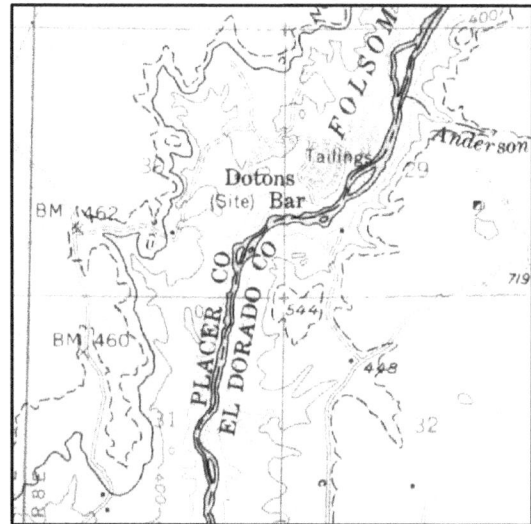

The dark solid line on this 1954 USGS topographical map is the North Fork Ditch. At BM 462 is the little bay of Beeks Bight when Folsom Lake is full. Note how far away the North Fork Ditch is from Dotons Bar on the map.

It's at the historic community of Dotons Bar (often referenced as Doton, Doten, and Dotan) that you can really see the devastation of gold mining along the North Fork of the American River when Folsom Lake hits 15% of capacity during a drought. Dotons Bar on the North Fork of the American River was heavily mined. Old maps indicated there were several buildings and a district known as China Town.

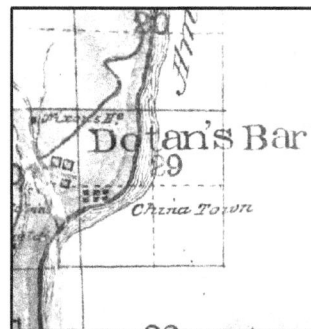

1887 Placer County map showing China Town at Dotan's Bar.

I did stumble across a broken rice bowl with Asian script (Chinese?) on the bottom.

Broken rice bowl with Asian marking on the base found amid the tailings at Dotons Bar.

Among the piles of river cobble that stretch for over half a mile was one lone pillar of unconsolidated mud flow standing like some sort of totem pole. Perhaps it was left standing as a marker between two different mining claims. But in a landscape where the river bank had been so ravaged by mining for placer gold, this 7-foot-tall vestige of ancient river bottom was very prominent.

32. Looking south toward Condemned Bar are the hundreds of acres of tailings from Dotons Bar mining operations.

There was no better illustration of the havoc wreaked by gold mining on the American River under Folsom Lake than at Dotons Bar. The river bank had been chewed away from the river and piles of river cobble left in its place.

33. A lone pillar of ancient river bottom that was not washed away by hydraulic mining at Dotons Bar. Over 6 feet in height, it gives a perspective of how much of the landscape along the North Fork was reduced by mining for gold.

In 1860, a short five years after the North Fork Ditch had been constructed; a correspondent on his way to the Alabaster Cave in El Dorado County noted the condition of the landscape along the North Fork of the American River.

Daily Alta California, Volume 12, Number 133, 13 May 1860

NORTH FORK DIGGINGS. A romantic ride of a mile and a half, directly south of the Franklin House, brought us to the brink of the North Fork of the American River. The road gradually winds around the bank, which is, in fact, but a gentle declivity to the water's edge. Here is another suspension bridge, on which we cross to the county of El Dorado. Just above the bridge, on the Placer county side of the stream, lies a luxuriant vineyard, the grapes of which rival, in

size and lusciousness, those of the far-famed Los Angeles. Patches of grain and vegetables also dot the hill-side, and watered by flumes which also, strange to relate, supply gold mines immediately contiguous thereto, with the aqueous element.

The whole surface of either bank of the river, as far as the eye can extend, is ruthlessly torn up, scarified and disfigured. In lieu of the green grass and beds of flowers which formerly decked the hill-sides, ugly piles of boulders and huge holes now mar the picture. The rockers and long-toms have long since disappeared from the then crystal stream, and its muddy waters now flow through the ungainly sluice-boxes, and plunge down the precipitous steeps in turbid torrents.

34. A rare remnant of riveted cast iron pipe that was used to carry water from the North Fork Ditch to the mining operations down by the river.

The Narrows

If you are not travelling along the North Fork Ditch as you travel north from Dotons Bar, you must traverse over increasingly steep slopes of sand. The hillsides, for so long drowned by the lake, are devoid of any vegetation and composed of virtually nothing but sand. The slopes are very steep, and the water laps at the sandy bank. One unusual feature is an outcropping of brilliant white rock that shines like a beacon in the afternoon sun. I'm unsure if this is an outcropping of pure quartz or limestone. If it is quartz, I often wonder why miners didn't attempt to chase this quartz vein for gold. Perhaps it wasn't really noticeable until the lake washed all the top soil and sand from around it.

35. Large exposed outcropping of white rock (quartz?) is like a beacon light next to the sandy hillsides when Folsom Lake is low.

The next significant bit of history that I found occurs at a spot that used to be called the Narrows, before Folsom Lake was formed. A large pluton of granite on the east side of the river pinches the North Fork of the American River at this point. Hence the references to The Narrows, because the canyon is very narrow at this spot. Today this point is named Anderson Island because it is surrounded by lake water when Folsom is full.

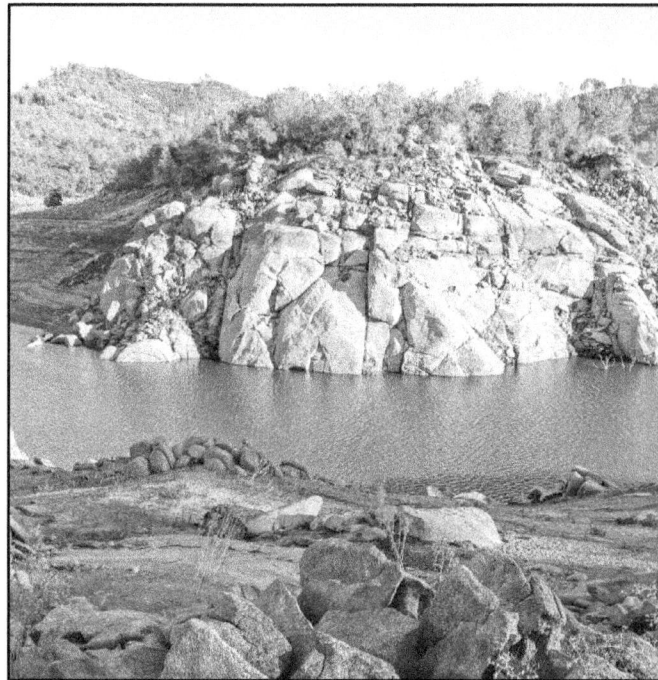

36. Walls of granite on either side of the river narrowed the river canyon in this spot south of Horseshoe Bar. Looking east toward Folsom Lake's Anderson Island.

The Narrows was also an ideal spot to cross the river with a bridge because you had granite bedrock on both sides of the river. On the eastern side of the river is El Dorado County, which has different geological formations from the western, or Placer County, side. There were many hard rock mining operations occurring in El Dorado County. One of these hard rock mines was the Zantgraf mine, which was located slightly north of the Narrows.

Narrows Suspension Bridge

When the lake drops below approximately 375 feet elevation, you will be able to see the stone masonry western abutment of a 19th century suspension bridge.

37. Western abutment for a suspension bridge that crossed over the Narrows supported a water pipe carrying water from the North Fork Ditch to mining operations on the eastern El Dorado County side of the river.

Chiseled into the granite is a shallow 10-inch-by-10-inch square, presumably to hold a vertical timber for the suspension cables.

38. Square-cut depression into granite for one of the suspension bridge's vertical wooden supports. A square iron nail I found rests at the top of the picture.

On the opposite side of the river, you'll be able to find suspension cable anchor points. Large holes were drilled into granite, and a solid piece of cast iron inserted to wrap the cable around. The suspension bridge carried a cast-iron pipe across the river.

4-inch diameter round cast-iron bars fitted into the granite anchored the suspension cable for the bridge on the eastern side of the river.

There is a large cut into the granite outcropping presumably to accommodate the cast-iron water pipe to travel east toward the mines. Judging from old pipe remnants scattered around the site, I estimate the water pipe to have been 12 to 18 inches in diameter. I have little doubt that this was the bridge that carried the water pipe from the North Fork Ditch over to the Zantgraf mine. The water may even have been used by other mining operations for hydraulic mining on the eastern bank of the river.

The granite was scalloped to cradle the suspension cable that supported the bridge.

The remnants of this suspension bridge should not be confused with the references to a suspension bridge across the river at Whiskey Bar. The nature of the landing on the south side of the river on Anderson Island does not lend itself for the crossing of horses or carriages.

Stone mason aqueduct

A very interesting structure a short distance north of the Narrows Bridge is the North Fork Ditch dry stone masonry aqueduct. However, it is difficult to see how this aqueduct or flume ever conveyed water, because it was not lined with concrete. Even though there is a channel on top of this stone structure, the gaps between the rocks would have prevented it from carrying any amount of water. In addition, the ditch actually loops around the stone viaduct, rendering it useless as an aqueduct.

39. In this picture, looking east, the waters of Folsom Lake have seeped through the stone aqueduct, creating a reflecting pool on the western side.

Nevertheless, the structure is fairly impressive. It is nearly 30 yards in length across the little gully and 15 feet from its base. It doesn't appear that the aqueduct was technically constructed by masons skilled in the art of stone rubble masonry, which employs the use of mortar in between the stones. Instead, it looks more like a dry stone masonry that uses minimal to no shaping of the rock surface before placement and no mortar to bind the rocks together.

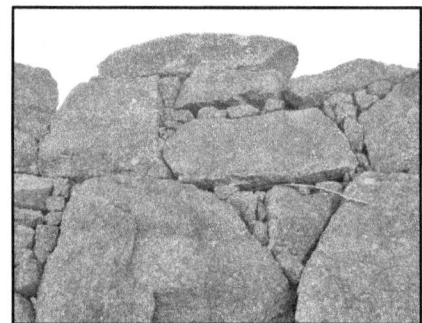

40. Careful attention was paid to filling all the gaps between the large granite rocks with smaller pieces for support and strength of the aqueduct.

The interstitial spaces between the larger rectangular granite blocks are filled with smaller stones. Its construction is more advanced than the rock retaining walls used to support many of the water canals in the area. But it doesn't rise to the level of the masonry used to build many of the wire suspension bridge abutments along the river.

41. A 3-foot-wide channel could have allowed water to be carried by the stone aqueduct, but today it is just a unique bridle or hiking path.

The North Fork Ditch did use wooden flumes to carry the water over some deep gullies and reduce the length of a canal dug out of the hillside. However, when the river would flood many of these flumes would wash out, requiring lengthy repairs. Perhaps the dry stone aqueduct was an experiment to see if the extra cost of construction provided better flume protection from flooding.

Horseshoe Bar to Auburn

Horseshoe Bar

Directly north of the Narrows the river flows around Wild Goose Flats, its flow resembling a horseshoe outline. As the river slowed down to wind around the landscape and pass through the Narrows, it deposited large quantities of placer gold. On my hikes this was also the location

where the flowing river met the still waters of Folsom Lake. Down by the shoreline the area is heavily silted and difficult to walk through. The silt and mud of the surrounding hillside have undoubtedly covered many of the building foundations that were once scattered along the river at Horseshoe Bar. I did come across what looked like a landing with a side foundation of river cobble rock.

42. A landing or foundation comprised of river cobble at the old community of Horseshoe Bar.

The North Fork Ditch had a particularly challenging stretch of landscape to overcome in this area. They had to essentially prop the canal up against a granite ledge. The subsequent concrete lining can still be seen, with local granite rocks supporting the canal on the hillside.

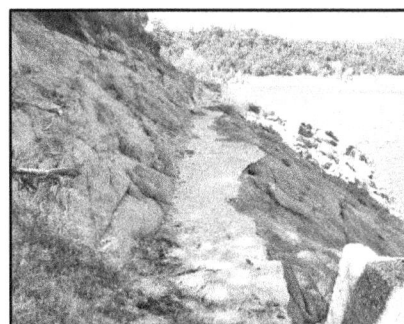

43. Remnants of the concrete lining of the North Fork Ditch cling to the side of a steep granite hillside above Horseshoe Bar.

44. Exposed riveted cast-iron pipe above Horseshoe Bar. Before the advent of welded steel pipe, sheets of metal were rolled and the edges riveted together to form links of pipe to carry water.

Whiskey Bar Bridge

The Whiskey Bar Bridge is mentioned numerous times in newspaper articles. As best as I can ascertain, it crossed the river just north of present-day Horseshoe Bar Road and landed on the western point of Wild Goose Flats. Unfortunately, I was never able to locate any remnants of this bridge, built in 1854.

Sacramento Daily Union, Volume 8, Number 1133, 9 November 1854

The wire suspension bridge just below this place [Rattlesnake Bar] is progressing very fast, the cables having been stretched across the river, and the approaches to the bridge on either side of the stream are nearly completed. The railway will be elevated far above the highest water mark, and the structure will be very beautiful and highly creditable to the taste and skill of the enterprising company who have erected it, and I hope that they will be well remunerated for their liberal outlay of capital on the road and bridge. Yours truly, Placer.

Daily Alta California, Volume 12, Number 147, 27 May 1860

An excerpt from a ***Trip to the Alabaster Cave***

Starting again, we left the Auburn Road, enjoyed beautiful views of several mining towns with euphonious names, and soon drove down a very steep hill to Whiskey Bar, where we crossed the north fork of the American River, on another bridge, which would be pretty but for the disfiguration of an ugly water pipe on one side. For picturesque beauty this neighborhood can hardly be excelled, embracing, as it does, vineyards and orchards and cultivated fields, flumes, sluices and ditches; tunnels and coyote holes.

At the top of the horseshoe bend in the river east of Whiskey Bar, I found more Native American grinding holes. You have to climb up and over some granite outcroppings to locate them. This would have been an incredible view as you looked down on an American River running wild and free. It is at Rattlesnake Bar State Park that the North Fork Ditch is higher in elevation than a full Folsom Lake at 466 feet in elevation.

Rattlesnake Bar

The community of Rattlesnake Bar was established in 1853 and suffered a devastating fire in 1864. There are old photos showing water cannons or hydraulic monitors dislodging the river bank for mining. At the eastern end of Rattlesnake Bar a suspension Bridge was constructed. It connected to roads from Cool and Mormon Island for the crossing of the river up to Auburn via Auburn-Folsom Road. While the bridge has become known as the Rattlesnake Bar Bridge, it was originally called Gwynn's Bridge for the owner. William Gwynn owned the

Alabaster Cave across the river from the mining community of Rattlesnake Bar.

The placer gold deposits were so promising throughout the Rattlesnake Bar mining district that a water canal was cut from the Bear River to bring water to the dry diggings in the region. The Bear River ditch was delivering water before the North Fork Ditch became operational in 1855.

Sacramento Daily Union, Volume 5, Number 694, 14 June 1853

Another Mining Enterprise. —The "dry diggings," as they are termed, will in a short time exist but in name. The last enterprise for metamorphosing them is the cutting of a ditch from the Bear river canal below Auburn, to Rattlesnake bar, some six miles below, on the North Fork. The object of the company is to furnish the immense flats contiguous to the bar with an ample supply of water. The work is progressing rapidly, and will be completed the latter part of next week.

<u>Gold Districts of California by William B. Clark*</u>

Rattlesnake Bar

Location and History. Rattlesnake Bar is in northwestern El Dorado county and southern Placer County. The placer mines here along the American River were highly productive during the gold rush. The town was established in 1849 and became good-sized until 1864, when it was destroyed by fire. The Zantgraf mine, the principal lode mine in the district with a reported production of $1 million, was active from 1880 to 1901 and again in the 1930's. Dragline dredging was done in the region during the 1930's. Part of the district is covered by the Folsom Reservoir.

Geology. The district is on the eastern flank of a major granodiorite stock that is intrusive into greenstones and amphibolite. A major body of serpentine and limestone lens crop out in the area. Several extensive deposits of Pleistocene shore gravels along the American River were hydraulicked. The Zantgraf vein contains abundant sulfides, including galena and chalcopyrite, and was mined to a depth of 1100 feet. The district also has yielded substantial amounts of chromite and limestone and some copper. (Clark, W.B. and Carlson, D.W., 1956, El Dorado County, Zantgraf mine: California Journal Mines and Geology, vol. 52, p. 429)

*Gold Districts of California, Bulletin 193 California Division of Mines and Geology 1976

Debris from hydraulic mining became problematic for communities downstream in the Sacramento Valley. A lawsuit was brought to restrict hydraulic mining in the 1880s. While upstream on the North Fork hydraulic mining was washing away entire mountainsides, the river banks beginning in the Rattlesnake Bar area were not as easily dislodged with water. From the court proceedings comes testimony about the Rattlesnake Bar mining community.

Sacramento Daily Union, Volume 14, Number 105, 21 December 1881

Edward Christy testimony

The tailings from Georgia Hill went into Devil's canyon, and those from Wisconsin Hill into Yankee Jim's ravine, and I think they both emptied into Shirt Tail canyon. The population of that locality at that time was very large. I was interested in a quartz mill in 1853 near Auburn — what was called Crouse Hill Mining Company. It was located in Baltimore ravine, two miles this side of Auburn, that was kind of a failure, and I left there in 1853 and went to Rattlesnake Bar. At that time there was very little mining going on at that bar. There was but one little ditch there then. The Bear River Company dug a branch ditch from Auburn to Rattlesnake, and I had the contract of bringing it in there, and they calculated that they would sell 600 inches a day. They sluiced into the North Fork of the American river. There were at that time seven big sluices running into the river there. There was some pretty hard gravel in this bar and it did not wash easily. It was a very solid material and had to be picked a good deal. There was a little surface loam, and the rest was red gravel. The mines on this bar were about 100 or 150 feet above the surface of the river. I mined at Rattlesnake until the North Fork Canal came in there. I moved down to Doten's Bar— that was in 1856. Rattlesnake at that time was quite a large place. There was a little theater at that time. Wells, Fargo & Co. had an office there. There were three pretty extensive hotels. There were perhaps 2,000 people at Rattlesnake and in the immediate vicinity. Rattlesnake was the distributing point for supplies for all the bars in that vicinity. The first hydraulic mining I ever saw was at Rattlesnake Bar, in the Colby claim. That was in 1853. The site of the first town at Rattlesnake Bar has all been washed away. It covered about fifty acres, and perhaps much more. I never measured it. Doten's Bar is from 100 to 150 feet above the river. The formation of this bar is similar to that of Rattlesnake Bar. The claim that I worked on Doten's Bar had a bank of fifteen feet. We gave the sluices eighteen inches grade and used very small heads of water.

What I find interesting is Christy's description of "pretty hard gravel in this bar and it did not wash easily." This concurs with the almost concrete-like landscape left behind at both Rattlesnake and Dotons Bar.

The hard cement-like nature of the ancient river bottom along the river could support tunnels and was difficult to mine for gold.

45. Native American grinding holes overlooking the North Fork of the American River at Rattlesnake Bar.

Rattlesnake Bar Bridge

By October 2015 Folsom Lake had receded to the point where the North Fork was flowing freely around Rattlesnake Bar. I took this opportunity to ford the river. The reward of fording the flowing North Fork, in addition to feeling a twinge of kindred spirit to earlier pioneers and gold miners, was the opportunity to explore the Wild Goose Flats area, in particular climbing over the southern abutment of the old

Rattlesnake Bar Bridge. Similar to the north-side abutment, a piece of the old suspension cable was still on site. The southern approach to the bridge was built on top of a granite out-cropping and does not have as much dry stone masonry as the north abutment.

An earlier wooden bridge across this spot on the North Fork, along with the Whiskey Bar Bridge, was swept away during the floods of 1861-1862. William Gwynn, owner of the Alabaster Cave and mining operation on the south side of the river in El Dorado County, rebuilt the bridge. Instead of the wooden construction, a wire rope suspension bridge was constructed using patented wire cable from Andrew S. Hallidie Company in San Francisco. The remaining bridge abutments are truly interesting examples of dry stone masonry, with the granite blocks carefully shaped and fitted.

46. Northern abutment of the Gwynn or Rattlesnake Bar Bridge.

47. Formed concrete rests atop the southern abutment of the Rattlesnake Bar Bridge, with a base comprised of granite blocks.

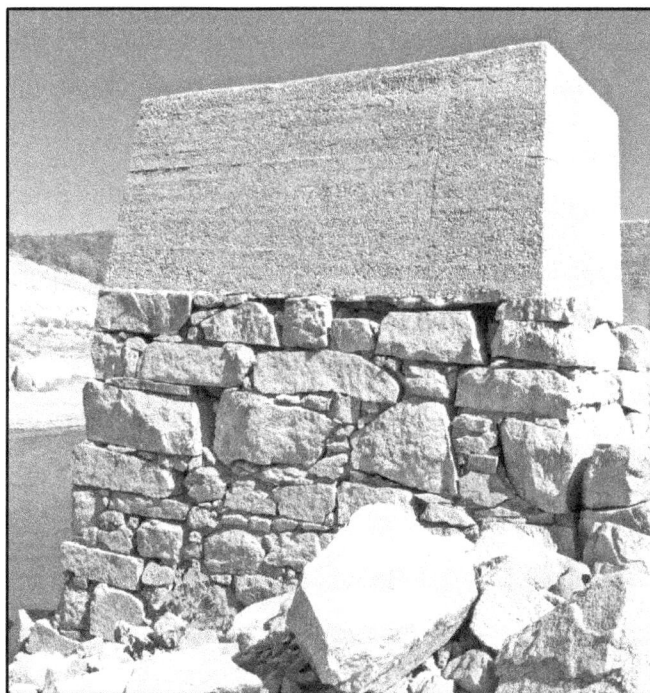

A young Walker Knauss standing on top of the north abutment of the Rattlesnake Bar Bridge looking southeast toward the southern abutment.

48. One of the vertical suspension cables is still present at the base of the southern abutment of the Rattlesnake Bar Bridge.

While the Rattlesnake Bar Bridge, which is listed as the Gwynn Suspension Bridge on an 1887 map of Placer County, was reinforced over the decades, it succumbed to an overloaded truck in 1954. Ironically, the bridge was slated to eventually be torn down, as it would be underwater with the construction of Folsom reservoir.

1861 – 1862 Floods

The rain began falling in December of 1861 and continued into January 1862. The ensuing floods not only inundated the City of Sacramento, but caused widespread devastation to the mining operations along the American River.

Sacramento Daily Union, Volume 22, Number 3344, 16 December 1861

THE INUNDATION IN THE INTERIOR.

From our interior exchanges, so far as the same have come to hand, we condense the following statement of the disasters by flood in the several counties during the week last past:

EL DORADO COUNTY.

The bridges at Coloma, Uniontown, Salmon Falls and Chile Bar, on the forks of the American river, were all badly damaged. No mail from Georgetown or Auburn had reached Placerville for about a week, on Friday. No Sacramento mail reached there during three days of last week.

The following in reference to the condition of travel in the county is from the Coloma Times:

Communication from Georgetown, Greenwood, and in fact from the entire ridge between the Middle and South Fork of the American river to Sacramento, can be had by the way of Shaw's wire bridge at Mormon Island, which, we understand, remains uninjured. There is but one bridge across the Middle Fork of the American river--that is known as the Murderer's Bar wire bridge. We have not heard authoritatively, yet we presume that it escaped--it is a wire suspension, far above high water, and if the abutments stood, no other injury could be done it. The Rattlesnake Bar wire bridge, we learn, was not damaged at all.

PLACER COUNTY.

The miners in Placer have suffered severely, and the damage to roads and bridges has been large. The Dutch Flat Enquirer says the Mineral Bar bridge, on the road from Illinoistown to Iowa Hill was carried away. This was considered one of the most substantial pieces of workmanship in the county. Ford's Bar bridge was also carried away. This bridge was for horsemen and foot passengers from Iowa Hill to Dutch Flat, and was one of the great thoroughfares from

Placerville, Georgetown, Forest Hill, Iowa Hill, Dutch Flat, Little York on to Nevada and Downieville.

The water reached its height on Monday. At Junction Bar, a wire-suspension foot bridge, several cabins, and a number of large waterwheels, flumes, etc., were carried away. Every house on Pleasant Bar was surrounded by water, but by means of ropes, stone ballast and other fastenings they were prevented from floating away. On Horse-Shoe Bar the damage was extensive, the place being nearly all covered with water. At this bar, and also at Mad Canon Bar, Poverty Bar, Maine Bar, and Oregon Bar, immense losses were sustained by the carrying away of water wheels, derricks, sluices, flumes, mining cabins, and lumber. At Volcano Bar, a wire-suspension bridge which cost $2,500 one year ago was destroyed. Large numbers of Chinamen are reported by the Courier to have been drowned at Poverty, Maine, and Oregon Bars, and at the confluence of the Middle and North Forks of the American.

Most likely the floods of 1861-1862 were a result of subtropical moisture streaming over the region, creating high rates of precipitation while at the same time the warm temperature association with moisture melted the Sierra snowpack.

Avery's Pond

North of Rattlesnake Bar the canyon opens up to Mormon's ravine. Above the river is Avery's Pond, in the ravine that was one of many holding and mud-settling reservoirs for the North Fork Ditch. One of the longest flumes of the canal spanned Mormon Creek directly north of the current PG&E Newcastle Power Plant. Manhattan Bar is north of the Mormon Ravine

entrance to the river, and the canyon really closes in at this point. The narrow canyon, which increases the velocity of a raging American River during times of high water, has erased evidence of the many historical mining activities in the area for which I could find records.

Mormon's Bar

I couldn't find Mormon's Bar specifically identified on any maps. It may have been at the entrance to Mormon Ravine. In any case, this communication to a Sacramento newspaper in 1852 represents some of the hostility and discrimination Chinese immigrants faced while gold mining on the American River.

Sacramento Daily Union May 5th, 1852

Mormon Bar, North Fork Am. River, Sunday, May 2d, 1852.

Messrs. Editors: — The excitement in regard to the Chinese is rapidly extending along the banks of the North Fork of the American River, and daily expulsions are taking place. This morning some sixty Americans ranged down the River some four miles, driving off two hundred, quietly removing their tents, strictly respecting their persons and property (except in one instance, when a Celestial seemed inclined to be "obstreperous," his cradle was thrown into the river. The same company intend to proceed "en masse" to Horse Shoe Bar this afternoon to concert measures with the miners there to "start" some four hundred located at that place. A band of music is engaged to accompany the expedition. The feeling is strong, and any thing but evanescent, that self-protection as a first law of nature, must and shall be enforced. Would it not be well for your city government to take some action on the subject, for you will certainly have a flood of them from above and below you soon. There is but one opinion among the miners in regard to the proposed monopolies and importation of Chinese into the mines, and nearly all of the eighty or ninety thousand American miners are fully determined to submit no longer to have the public lands robbed of their only treasure. Yours in haste, as I accompany the expedition down the River. Chadbourne. [COMMUNICATED.]

While hostility directed at Chinese immigrants may have erupted in 1852, subsequent newspaper reports indicate white miners worked their claims alongside the Chinese in later years.

Manhattan, Oregon, Little Rattlesnake, Tamaroo Bars

Numerous mining bars are referenced in newspaper articles. But north of Rattlesnake Bar few of the many mining bars have survived on maps. This may have occurred because no communities or buildings survived after the miners left their claims. As the river canyon narrows, there just aren't many suitable places to establish towns. The main road up to Auburn that was travelled by the stage coaches is several miles to the east of the river. The mining bars listed on many maps include (from directly north of Rattlesnake) Manhattan, Oregon, Little Rattlesnake and Tamaroo Bars.

49. Well-preserved weir outlet of the North Fork Ditch above Mormon Bar. This part of the North Fork Ditch is above the high-water level of Folsom Lake.

The hills on either side of North Fork of the American River get progressively taller, and the canyon gets proportionally narrower as you hike north of Mormon Ravine where the Mormon Creek flows into the North Fork. Parts of the canyon where the river flows are reduced to 100 yards across, with steep bluffs on either side. Where the canyon does open up, some of the river bars that were being worked were only 150 yards wide by 500 yards in length. There are no bench placers in this part of the canyon. It is a tough hike in this part of the river because there are no established paths, and you are walking over river cobble rocks. But this didn't stop the miners from diverting the river so they could get to the gold sitting on the river bedrock.

50. The rock retaining wall supporting the North Fork Ditch has fallen away, exposing the underside of the concrete lining installed in the 1910s.

This correspondence from 1858 illustrates how miners were diverting the flow of the American River in order to reach the streambed in search of gold.

Daily Alta California, Volume 10, Number 252, 13 September 1858

MINING ITEMS

A correspondent of the Sacramento Union, writing from Auburn, September 9th, gives the following notice of mining operations in that vicinity:

Having a little leisure time yesterday, I took a stroll down the river to witness some of the mining operations that are carried on near this place. The first claim I came to was that of Harper & Co., on Little Rattlesnake Bar, one mile from town. This claim belongs to a company of five, and was located in 1850, and has been worked every year since with one exception, in 1853. It has always paid the owners well. They have a flume five hundred feet long and eighteen wide, which carries all the water of the stream. The dirt is taken up in cars that are drawn by a wheel drawn across the flume. The company commenced washing last week, and made $6 per day to the hand in top dirt, and expect to find it rich when they reach the rock. Seventeen men are at work day and night on this claim.

Just below is a company of Chinamen, working by means of a wing dam and race cut. I could not learn from them how they were doing. Next below, and in sight of Harper & Co., is another set of Chinamen at work. They have a flume six hundred feet long and are fixed for working very well. They have an overshot wheel for pumping out their claim, and use water from the American river ditch. The same water that turns their wheel also washes the dirt. There are several Chinamen at work here and doing very well.

Passing upstream a quarter of a mile, brings you to the Empire Company, owned by Anson, Murphy & Co. Their flume is the same length and width of Harper & Co.'s. There are four shares in this claim, and the company are only working one extra hand. They have not much dirt to wash, owing to the bedrock rising so high. They are making good wages washing every other day. They will work their ground all out by the time the high water comes in the Fall.

Next above the Empire is the Lone Star Company, owned this year by a company of Chinamen. Their flume is 400 feet long. They have not drained it as yet. The claim was owned and worked, last year, by Murphy & Co., who did well, and sold out in the Fall for a good price, and took up the one below, where they are at work now. This Company expects to do well as soon as they get it drained.

Next above is the Tamaroo Claim, owned by McDaniel, Thurman & Co. There are six shares in this Company. The claim was located in 1850, and has been worked every year since, and always paid well until last year, when they were driven out just as they were getting to good pay. They have about thirty hands at work; one-half of whom work in the day time, and the others at night. This Company are working with a wing dam, and washing pay dirt. They have their claim well opened, and it bids fair to reward the owners for their labor and expense. At the upper end of this claim the American River Ditch [North Fork Ditch] is taken out, which runs on down, - supplying Rattlesnake, Doten and Beal's Bars with water, terminating on the plains somewhere near Folsom. It is a pretty flume where it is taken out, being about five feet wide and three feet deep. Miller, the agent here, says it runs full all the year, and pays the Company a handsome dividend monthly.

Similar to the scouring of the canyon inflicted by the floods of 1861 and 1862, successive high river flows have erased all but the most intransigent artifacts of mining history. The remnants left behind are usually the rock retaining walls and concrete structures of the North Fork Ditch.

51. A multi-weir concrete structure of the North Fork Ditch has withstood the ravages of flood waters in the canyon since the 1950s.

52. A view on top of the multi-weir concrete structure of the North Fork Ditch looking south. Note the steep slopes of the hills of the river canyon.

Birdsall Dam

Located at the upper limits of the high-water level of Folsom Lake is the Birdsall Dam location. Approximately 3.5 miles north of

Manhattan Bar, the Birdsall Dam was the last of several dams constructed in the 19th century on the American River that diverted water into the North Fork Ditch. The first dam for the North Fork Ditch was constructed across Tamaroo Bar. As a reference point, Tamaroo Bar is approximately one mile north of where the Auburn Dam was to be constructed. The Auburn Dam was situated across Little Rattlesnake Bar.

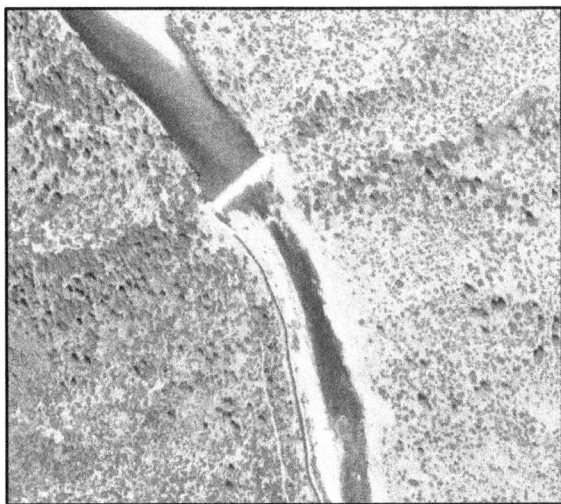

1952 aerial photograph of the Birdsall Dam diverting water into the North Fork Ditch.

The original Birdsall Dam was constructed in 1880. In 1898 it was enlarged with a new rock and masonry downstream face. It spanned across Poverty Bar downstream of Little Rattlesnake Bar. The final iteration of the dam was a sturdy structure that anchored the eastern end to hillside bedrock with iron tie rods. There are still some rock and concrete remnants on the west side of the river and the anchors set into granite bedrock on the east side.

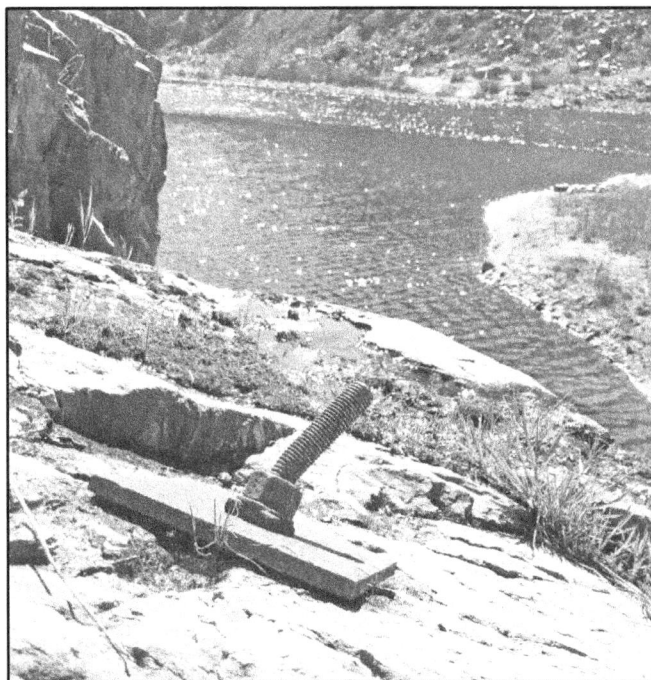

53. Anchor bolts and plates that secured the Birdsall Dam to the east side of the river canyon still remain.

Interestingly, the Birdsall Dam was situated right at the outlet of the American River diversion tunnel built to divert the river around the Auburn Dam construction site. The diversion tunnel has now been sealed off with rocks so you can hike over to the anchor points of the Birdsall. Technically, the high water of Folsom Lake is right below the old dam and at the foot of what would have been the Auburn Dam, had it been built.

54. Anchor bolts drilled into solid rock helped anchor the eastern end of the Birdsall Dam.

What is confusing is the location of the Birdsall Dam on the 1910 American River and Mining Water Canal Map, which places its location at the original Tamaroo Bar site. The 1954 USGS Auburn topographical map places the Birdsall Diversion Dam in Section 23 of Township 12 North, Range 8 East. But the 1910 water canal map mirrors an 1887 map of Placer County that locates the dam upstream in Section 14 of Township 11 North, Range 8 East.

The Birdsall Dam and connecting North Fork Ditch were responsible for supplying water to the northeastern farms and suburbs of Sacramento County. By the early 20th century clean drinking water was more important than gold, in the eyes of suburbanites.

Sacramento Union, Number 32, 2 October 1914

Miner Arrested for Polluting Water

AUBURN (Placer Co.), Oct. l.—J. A. Anderson of the Pacific Gold Dredging company, operating at Mammoth Bar on the American river east of Auburn, has been arrested and charged by Sacramento parties of polluting the water, which is later taken up by the North Fork Ditch company and sold for drinking and Irrigating purposes between Auburn and Fair Oaks. At this time of year, when there is no mining above Auburn on the north and middle forks of the American, the water in the north fork ditch is usually as clear as crystal. It is always muddy in the winter time.

Knickerbocker Waterfalls

The Knickerbocker waterfalls have nothing to do with the history of Folsom Lake. But they are spectacular, if you can ever get close enough to see them. Located a quarter mile downstream of the Birdsall Dam site, Knickerbocker Creek spectacularly crashes through the canyon it has cut on the east side of the river. From the trails on the west side of the river you can see the waterfalls during the winter and spring months when lots of water is flowing down Knickerbocker Creek.

55. *One of the many waterfalls of Knickerbocker Creek that discharges into the North Fork of the American River from El Dorado County.*

While I have hiked down into Knickerbocker Creek Canyon to capture some photos of the waterfalls, I would not recommend it because of the steep terrain, poison oak and ticks—both of the latter which I acquired climbing out of the canyon.

56. *Knickerbocker Creek has carved what looks like a water park amusement slide into the hillside rock.*

Wild Goose Flats

El Dorado County, on the east side of the North Fork of the American River, is not conducive for hiking. The river usually runs right along the edge of the steep hills that form the lower part of the American River canyon. When the North Fork was running particularly low in the summer of 2015, I forded the river from Rattlesnake Bar to the south side known as Wild Goose Flats. The water wasn't particularly cold and was only calf-deep in most places. As I waded through the rushing river water, I was careful to place my next step only where I could see a rock bottom. I wondered if men and women 150 years ago felt the same sense of adventure and trepidation crossing the river on foot as I did.

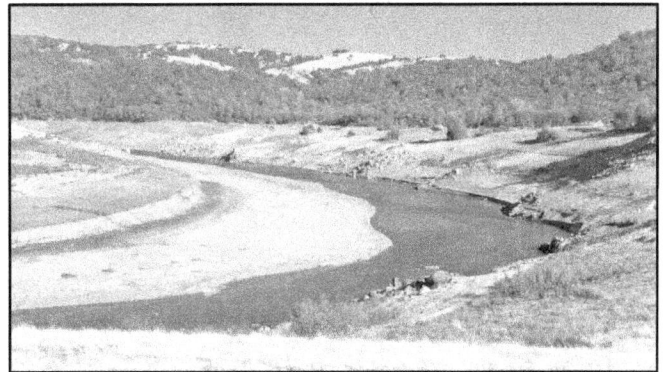

57. *North Fork of the American River runs free around Rattlesnake Bar as viewed from Wild Goose Flats, looking northeast.*

River Gauging Station

The first historical feature present on the north side of Wild Goose Flats is the Rattlesnake Bar bridge abutment. West of the old bridge next to the river is a concrete structure, tall and wide enough for a person to stand inside. The 1944 Auburn USGS topographical map indicates this was a river gauging station. I'm not sure how it worked, if it was monitored with electrical gauges, but the hillside is relatively steep on the south side of the station. A significant amount of sand and silt has built up around the gauging station, making it difficult to climb around as the river runs right up against the river bank.

other part on the river; that is, there are some claims that pay better, and all that work are making wages. Rattlesnake and Whiskey Bar have been known since '49 as a good locality for gold, and the Bear River Ditch has of late sent it water.

Goose Flat is a new place and small at that. There are half a dozen claims that pay large from an ounce to three ounces to the man, and the few others wages. This place has an elevated site, which has prevented its being prospected earlier, for it is forty feet above the North Fork Ditch, and thirty above the nearest point of Bear River Ditch. The water they have has been in but few weeks, but will last till August and furnish a supply eight months in the year. Were the grounds more extensive an effort might be made to bring in water from one of the forks of the river, but it has been decided an unpaying enterprise, as the water is needed only a small part of the year.

58. Concrete river gauging station opposite Rattlesnake Bar on the eastern side of the river.

Parts of the Wild Goose Flats area are above Folsom Lake. There are a variety of roads and trails leading to various placer mining operations. The area is still accessible by a private road off of Rattlesnake Bar Road.

Sacramento Daily Union, Volume 8, Number 1172, 25 December 1854

But I must tell you of the towns further up the river — of Rattlesnake Bar and Wild Goose Flat just opposite, and Whiskey Bar below it. The mines are all richer in these places than in any

59. One of several concrete foundations in the Wild Goose Flats area that might have been part of a mining operation.

Further south across from Horseshoe Bar are the remnants of river bank placer mining and the resultant river cobble tailings. South of these tailings and up the hill, I came across more Native American grinding holes.

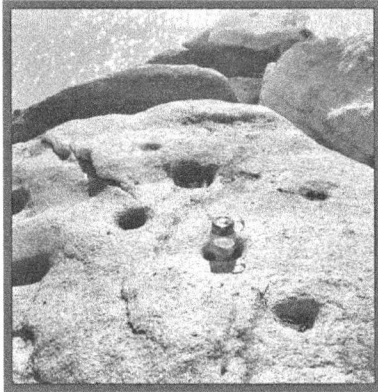

60. Native American grinding holes in the Wild Goose Flats area.

61. Parts and pieces of what looked like a wood burning stove.

62. Concrete foundation west of the entrance of the Zantgraf Mine. It is only 2 x 4 feet in dimensions.

Zantgraf Mine

If you hike south, before you reach the east side of Anderson Island that creates the Narrows where the suspension bridge was located, there will be a ravine up into the eastern hills. Up this ravine, over piles of mine tailings, is the Zantgraf

Mine. It is above the Folsom Lake high waterline. I came across some concrete foundations that looked like they were for pumps. The water pipe that crossed the river on the suspension bridge traveled over to the Zantgraf, so inevitably there should be some remaining supports for the water pipe or pumps to push the water up the hill.

63. Tall stone rubble masonry structure, possibly part of the stamp mill, for the Zantgraf Mine. The dark areas are moss.

It was reported on July 24, 1897, in the Placer Herald that the Zantgraf mine had installed a bridge over the American River to support a pipe to deliver water from the North Fork Ditch over to its operation. It also noted that the water pipe, "hung by cables across the river," was north of a power house. Antone Zantgraf, along with his brother Jacob Zentgraf, emigrated from Germany and were skilled in stone masonry. (There are several different spellings of the last name Zantgraf. Brothers Jacob Zentgraf and Antone Zantgraf had their last names spelled differently.

Maps usually referred to the Zantgraf Mine, while newspaper articles use Zentgraf.)

64. Concrete foundation with radial slot in the middle, most likely for the water wheel that drove the stamp mill at the Zantgraf Mine.

Daily Alta California, Volume 81, Number 5, 5 July 1889

One of the steady-paying quartz mines on this divide is the Zentgraf mine at Wild Goose Flat. Thos. Stevenson, who called at the Gazette office Sunday, informs us that the new mill for this mine will be ready to start up early in August. The new mill is being built just below the mouth of the new tunnel, which opens the ledge 116 feet deeper than the old works, or about 300 feet below the surface. The mill will be run by a hurdy-gurdy wheel, with power from the California Water Company. The old mill was run by an overshot wheel. The ledge varies in width from three to five feet, and pays for a distance of some 700 feet in length. This will keep the mill busy for several years. The mine has been steadily worked for the last eight years, paying the owners handsome dividends. The ore yields $6 or $7 a ton, if we remember rightly.

In 1899 the Zantgraf Mine announced an expansion of its mining operations. Operators added a boarding house for the employees and five stamps to the mill. In February 1899 there was a fundraiser in the City of Folsom to raise money for a new road to the Zantgraf Mine. It was reported in March 1899 that a contractor by the name of Derrington was to have the new road finished in April. After the illustrious start to the new year, the fortunes of the Zantgraf seemed to unravel. It was during this time period that the Zantgraf was sold to a mining company from the East Coast.

In November of 1899 a young mine employee by the name of Bartlett Irving travelled to San Francisco, where he got a hotel room. He borrowed a pistol from an acquaintance and committed suicide by shooting himself in the chest. The story about Irving's suicide noted that he had practiced law in San Francisco. No note or further details as to why he committed suicide were revealed.

The following year in December, the mine had to be shut down because a canal supplying the mine with water broke. It is unclear if the broken water canal was the North Fork Ditch or the Georgetown Ditch, which also supplied water to the mine. In 1901 a miner was killed when a load of timbers broke free in the mine and pinned him against the mine shaft wall.

Newspapers reported in September of 1901 that an Eastern company (the Montauk Consolidated Mining Company) was preparing to reopen the Zantgraf. The following month in October, the mine was hit by a fire.

San Francisco Call, Volume 90, Number 137, 15 October 1901

Fire Destroys Mine Works

AUBURN, Oct. 14.— The Zantgraf Mine Works, located six miles from here on the El Dorado County side, was burned this morning. The loss is about $30,000. The fire caught in the carpenter shop, destroying the mill, hoisting works and a portion of the shaft. The miners escaped through a tunnel at the 300-foot level. The Zantgraf has been a great gold producer.

It would later be revealed in a lawsuit that, according to the Sacramento Union in 1908, "In April, 1901, while in debt more than $20,000 for borrowed money and on other accounts, the company made a deed to its President, William Dallas Goodwin, who agreed to pay the debts, and claimed to be a creditor himself for a large amount, but instead of paying off the indebtedness he made certain preferences, according to the complaint in the case, whereupon California creditors filed a petition in the Superior Court of El Dorado County to have the company adjudged insolvent under the laws of California." The Superior Court of El Dorado County cancelled the deed, and the assets returned to the original company.

65. Stone rubble retaining wall in the small ravine below the Zantgraf Mine operations.

The portion of the Zantgraf that I was able to visit looked like part of the old stamp mill. The Zantgraf was a hard-rock mine bringing gold-bearing ore up from a mine shaft to be crushed. The tall stone-rubble edifice is over 30 feet tall. There is a concrete platform with a radial cut in the middle for some sort of wheel. The structure is wider at the base than the top and is constructed of local stone and mortar. It's interesting to contrast this tall structure to other masonry construction in the area.

Unlike the North Fork Ditch aqueduct and Rattlesnake Bar Bridge abutments, this Zantgraf main structure looks to have used irregularly shaped local stone bound with mortar. The former were constructed from carefully selected, and possibly minimally shaped, pieces of granite layered horizontally. Like the bridge abutments, the Zantgraf structure tapers from the base to the top. Below the mining structure is a rock retaining wall constructed of stacked stones with no mortar.

Anderson Island

When the lake elevation is above 430 feet, the hill that creates the Narrows on the North Fork of the American River becomes an island. Surrounded by lake water when Folsom Lake is full, this hill becomes Anderson Island. With the lake level low, you can hike the cow trails between Anderson Island and the higher hills to the east. There were old reports that a hydro-electrical plant was located on the west side of the river south of the suspension bridge. Presumably, water from the North Fork Ditch was released to spin a generator near the river's edge. I searched high and low for any remnant of such an operation. All I was able to stumble across were concrete pilings with angle iron sticking out of the tops directly to the east of Anderson Island.

66. Concrete piling that supported 2-inch angle iron south of the Zantgraf Mine. There is another piling approximately 15 feet to the north. This is looking north, just east of Anderson Island.

Were these concrete pilings for electrical utility poles? I'm not sure, but they are positioned correctly to carry electrical wires to the Zantgraf mine right over the hill. Given that, by some accounts, an electrical generator was located across the river, it's possible that electrical lines were strung across the river to the Zantgraf.

Peninsula

You can hike from Anderson Island down to the Folsom State Park Peninsula Campground, but I would not advise it. Below the high-water line of

Folsom Lake, the hillsides are steep and rocky, dropping right off into the lake. Above the high-water mark is dense oak savannah woodland. An old trail from the campground was abandoned years ago and is overgrown. I've hiked along the eastern side of Folsom Lake to the Peninsula Campground, and it was challenging, to say the least.

67. Concrete bridge across a small creek north of the Peninsula Campground east of Condemned Bar, normally under the water of Folsom Lake.

When Folsom lake is at normal levels, the high ground between the North Fork and the South Fork of the American River looks like a peninsula cleaving the lake surface in the middle and pointing at Folsom Dam. Folsom Lake State Recreation Area developed the Peninsula Campground, which is not a primitive site. There are lots of campsites, well maintained roads, and multiple cinderblock bathrooms. However, it is difficult to get to on a winding Rattlesnake Bar Road, nine miles from Highway 49 south of Cool, CA.

68. Native American grinding holes might have been in use on the Peninsula when the early gold miners set up camp in search of gold.

Consequently, this park and campground are never really crowded. Once you get to the Peninsula Campground, there are miles of wide trails and dirt roads for hiking and biking. Part of the Peninsula Trail is the old Peninsula Road from Mormon Island, and you can still see remnants of the asphalt roadway on some sections. Early reports called this road from Mormon Island the Georgetown Road.

But before there was a name for this path, there were early miners, fresh from a trip across the plains, who trudged over the trail in search of gold.

Experiences of a Forty-Niner

It was an old book titled "Experiences of a Forty-Niner" by William G. Johnston published in 1892 that first captured my imagination about the peninsula area. Johnston was on one of the first wagon trains to enter California during the gold rush in 1849. Forty years after his experience he penned a book about his gold rush travels and adventures. Specifically, he devotes a chapter to mining on the North Fork of the American River. It is a straightforward reminiscence with little hyperbole or embellishment as far as I can reckon.

Johnston gives a fairly detailed account of crossing the South Fork of the American River to travel to a mining camp on the North Fork of the American River. It is one of the best descriptions of life and death I have read of the early days of gold prospecting on the American River.

Google Play Books starting at page 270, Chapter XVII, Gold Mining – Our North Fork Camp

Having a wagon engaged for transporting our goods to Mormon Island, (South Fork of the American River) twenty-eight miles distant, we made an early start. Nine miles from Sacramento we halted to dine; and eleven miles further, at Oak Springs, camped overnight. A march of a few hours on the following morning brought us to Mormon Island, and we pitched our tent on a hill-side beside the South Fork, intending to remain there until we could ascertain the whereabouts of the friends we sought. Our goods weighed scarce a ton, and for wagoning to this point we paid seventy-two dollars.

After much inquiry, we learned from a storekeeper who claimed as his customers those we were in search of, that he usually saw one or the other about once every week, but did not know where they were located.

Mr. Reppert had contracted dysentery when on his sea voyage, and the disease still clung to him, and confined him closely to the tent; poor fellow, he never regained health. A temporary illness made me his companion for a few days.

Meanwhile Messrs. Murphy and Barclay did some "prospecting," but without making discoveries of any importance. Before that I was able to go down to the river where all day long the miners were rocking their cradles, I became very curious to witness the operation. From sunrise until dusk, I could hear the machines as they swayed to and fro, beating against the rocks on which they stood and filling the air with their singular music; while mingling with these sounds was that of the peculiar rattling caused by gravel descending to and shaking in the sieves. I felt too, a desire to know the actual results of such labors, and to see the glittering stuff, in the gathering of which these sounds were produced.

Two days later, Mr. McBride happened to cross our path when on his way to pay a doctor who had attended him in a recent illness. Fever and ague, from which he had recovered, had reduced him greatly, but what was left of him, we were all rejoiced to see — his heart was as big as ever. Mr. Scully and he had located on the North Fork of the American River, three miles distant, and separated from the branch on which we were by a high hill. They had at first worked at this place, but were induced to go to the other stream by the representations of our late overland companions, "Old" Smith and Eli Nichols, who were located there.

It needed no formal invitation to unite forces, that was a foregone conclusion on both sides; and without stipulation of any kind, all we had was in common, with no reservations. We at once procured a skiff and crossed with our goods to the opposite side of the river. From the landing to the base of the hill was but a few rods, and thither we conveyed our stores, piling them up carefully and covering them sufficiently to protect them from the weather; setting aside however, what we designed carrying with us to

our intended camping ground, concluding to return from time to time for such fresh supplies as we might require. Although this cairn was in sight of hundreds of miners on Mormon Island, we felt no concern as to its safety, such was the confidence then felt as to the respect for law written on the consciences of men, for here as elsewhere through California in general, there was a total absence of both civil magistrates and of civil laws. No provision under authority of the United States had as yet been made, and it was a disputed point as to the prevalence of Mexican laws.

The miners had a code of their own, and if any one outraged the decencies of life, they either drove him off, or hung him; these were the common penalties, and a few examples caused them to be well understood.

Each one started with a load strapped to his back or otherwise carrying what he could; and by a path leading up the steep hill-side, we ascended leisurely until the summit was reached, after which the way was almost level. There was an unbroken forest of stately trees, mostly pines, the entire distance. An hour's walk brought us in sight of the valley of the North Fork, when by a precipitous path, we descended to the camp of our friends. Mr. Scully was absent, and we made the rocks and hills ring with our united voices shouting halloo! when he at length came forth from a cleft where he had been prospecting. Wearing a red flannel shirt, pants rolled to the top of his boots, his head sheltered by a great felt hat, with bushy whiskers almost covering his face, and carrying a pick, shovel and pan, he was the very personification of a California miner!

After the hearty greetings which transpired in a universal chorus, we were introduced to an

English sailor whom he named as "Dick," whose tent was but a short way off, and who also appeared on the scene attracted by our shouting. Dick hailed from Melbourne; was industrious, quiet and thoroughly companionable. Smith and Nichols were encamped about a mile further up the stream; and about a mile below, there was a company of men from the island of Manilla. So far as we knew, there were no others on this stream, at least there were none near at hand. It was quite an advantage to be thus situated, as we were at liberty to move about at pleasure, without being compelled to "stake off a claim," as was usual in most mining districts.

The tent we brought with us was sufficiently large for the newly formed mess, and we pitched it near to the base of the hill under the shade of two wide spreading oaks. The wagon cover which had been used as a substitute for a tent by our friends, we employed as a roof for our dining room, stretching it upon posts. A goodly camp fire is ever an essential for camp life; and Mr. Barclay's sinewy arms applied to a sharp axe, were not long in felling a great pine tree, and from its trunk was severed a large back log, and another to sit upon, while very soon a splendid fire, hissing, cracking, sending up its volume of smoke and showering its sparks around, contributed to our cheer. We cooked a pan of "'rashers" with "hard tack" soaked in the fat; these and a pot of coffee added, formed our evening meal. With splendid appetites we desired nothing more, and felt that no king fared better. "The cloth being removed," we gathered around the fire, filled our tobacco pipes, and as the smoke curled upward from them, in story we "fought our battles over again." The plains were recrossed, and the Horn redoubled — and it might be other horns were doubled, perhaps trebled.

After a good night's rest and an early breakfast we went to work — this was on Friday, September 14th. The site chosen for digging was scarce more than a stone's throw from our tent, and we moved but slightly from it during our stay in the mines. From this point the earth had to be carried in sacks to the edge of the river, where it was emptied into two cradles for washing. It was a rough, precipitous descent of rock, twenty feet or more in extent, over which this carrying was done. The earth in which gold was found in this vicinity lay to the depth of from four to six feet upon the top of rocks. Usually we cast aside as worthless, a layer of sand, or sometimes sand mixed with earth; all below this was carried to be washed, and the nearer we approached the rock, the greater were the returns. When the surface was reached, it was scraped and swept with great care, for here the gold could be seen, lying in shining particles. North' Fork gold was uniformly composed of diminutive scales or mere specks, all of the utmost purity. In digging the earth with a pick, it was broken up fine before being shoveled into bags, and as many stones as possible were picked out and thrown away ; even then in washing it was found that a great part of the refuse was gravel, and this often quite large.

Mr. Reppert was continuously confined to the tent, or close to it, and of course unable to work. Two of us did the digging, one carried the earth and two worked at the cradles. At the outset I tried rocking a cradle, thinking a knowledge of that kind might someday become useful. It seemed at first sight quite simple, but on trial I found it rather difficult to acquire the peculiar art of moving one arm forward and backward to do the rocking, and with the other to dip up water with which to drench the dirt in the hopper. Skill too was required to know the exact motion that would produce the best results. If shaken too violently, the fine gold would pass out over the cleats in the bottom of the cradle and be lost; and if not swayed rapidly enough, the dirt would become packed behind the cleats. The miner too, required to learn his machine, as the height and curve of the rockers, the size of the cleats, and the slant of the cradle towards the stream, had much to do with proper working. I was so much exposed to water, my feet constantly in puddles, and my clothes dripping with wet, that I was glad to exchange cradle work for what was usually accounted severer Labor — digging with a pick. Messrs. Murphy and Barclay were broad shouldered, stalwart men, and one or the other, or both, at times carried the bags of dirt, which was indeed laborious work, requiring too a firm step and great care in descending the ledge of rocks to the river.

A full bushel of earth mixed with stones was the load carried, and frequently seventy-two bushels were washed in a day. Our days, however, were short, i.e., our working hours in each day, and at no time did we overtask ourselves. We mingled recreation with work, finding amusement in fishing and hunting. In angling we were never very successful, but as the forest on the hill above was full of game, our table at all times was kept well supplied.

Messrs. Barclay and Murphy, especially the former, were excellent sportsmen, and it was rarely ever that we were without a deer, or several hares, or jack rabbits as they were commonly called, hanging from the limbs of trees near our tent. I never but once went hunting, and then speedily enough tracked a young deer, and got within easy range of it, but was completely overcome with buck fever, being wholly unable to raise my rifle, and so enchanted

with the beautiful sight of the leaping of that deer from point to point — down into a ravine and then up a hill-side opposite, that I gazed at it with stupid admiration, until it was clear out of sight!

We usually kept the hares hanging in the open air until they became quite tender and blown by bees and wasps. This was said to be French style, and in our cuisine we were close imitators of the French. While preparing these animals for cooking when thus cured, there might be some offense to delicate nostrils, but when once cooked and spread out on the table they were considered very savory. This digression about hunting and mining work, is but a practical illustration of our manner of life; some work and some play; much, it must be confessed, of the latter.

I should have added that in washing the earth in cradles, a heavy black sand remains deposited behind the cleats, and at intervals this was taken out, and for a time kept carefully in a vessel used for the purpose; for it was in this sand that the gold was found. It was usually the work of either Mr. Scully or Mr. McBride, to separate the gold from the sand. At their convenience they panned it off — that is, seating themselves by the river's edge, and having the sand in the tin pan, they would so tilt it as to admit water, and again tilt it to allow the water to pass over, carrying sand with it. This operation was repeated with extreme care a great many times until the gold was left nearly clean. What remained was next spread upon a plate and placed in the sun to dry, when by a skillful use of breath the sand would be blown off, leaving only the gold, which was then stored in a buckskin bag.

We sometimes tried other "diggings," but never with as much success as in the place originally selected. Once we crossed to the opposite side of the stream, where the rocks bordering the river were higher and more precipitous, but were not repaid for our labor. At times we climbed among the rocks, and with knives or spoons picked out the dirt found in crevices; but while this earth was often quite rich, it took so much time to gather it, that the conclusion reached was that this too was unprofitable.

On the plains, the Sabbath with us, and with most emigrants had been as another day, and we scarcely knew of its arrival, having become careless as to the obligation to refrain from labor. Personally, however, I ever knew of the return of the sacred day, and was aware too that I was acting in disregard of a father's express counsel, given me before starting — to do no work on the Sabbath. "Sunday work" said he, "never prospers." I always felt, however, that it would have been utterly useless for me even to raise my protest. In the mines I do not remember that it was ever suggested to rest on that day; but by silent, common consent we refrained from labor as we had all been accustomed in our far off homes.

The average daily earnings per man were about one-half ounce of gold, or say eight dollars; once I remember we made double that amount. All worked in perfect harmony, and at no time was ever a word uttered to mar the pleasant relations existing. There was no shirking of work on the part of any one, and indeed if the "heavy end of a log" had to be carried, there was a generous rivalry to secure that end.

There was, however, one cloud in our skies; our Baltimore friend, Mr. Reppert, was a confirmed invalid, and we saw no prospect of his recovery. A physician at Mormon Island was consulted from time to time and the medicines he furnished

were given with regularity, but all seemed unavailing. Indeed we felt little or no confidence in this medical man, having the belief that he was a quack. Mr. Reppert had been repeatedly urged at a time when he was able to do so, to return to Sacramento, where he might obtain proper medical treatment; but while again and again assenting, he postponed doing what was suggested until wholly unable to go. He died Sunday, the 18th of November. On the day following, his remains were borne by our neighbors the Manillanos, who kindly volunteered their services, to the brow of the hill immediately back of our camp, and were there deposited in a grave which had been dug by his sorrowing associates. From the spot where his body reposes the beautiful North Fork may be seen winding about in the valley below, amid the grandest of scenery; but what careth the dead for this? There is no pride of place in the grave.

Mr. Murphy and myself were absent at Sacramento at the time the sad event just recorded took place. We had left the mines about a week previous to Mr. Reppert's death, little thinking his end was so near at hand. The occasion of our leaving was this. The rainy season had set in: the first rain since our entrance into California, had fallen on the 10th of October; after that there had only been occasional showers, until the 1st of November, when it began to descend in earnest, and continued with little intermission until the 7th of that month.

Confined to our tent most of this time, we saw plainly that preparations must be made for the winter, or wet season, as it would be impossible to remain much longer in quarters so unsuitable. On the question of wintering where we were, we were not at first of one mind, for it was feared by some that on account of the deep snows we would seldom be able to do any work; on the other hand, it was argued, that if we went to the cities, the expense of living would be such, that when spring came we would be moneyless.

On the night of November 9th, after much debate it was finally concluded to set to work on the day following to build a cabin of logs to winter in; and when our candle was blown out, and each one had sunk into quiet slumbers, if there were dreams, it might be readily imagined the subject of them would be of forests bowing to the axes of sturdy woodsmen — of a log-cabin rising grandly on the rocky shore of the North Fork, with coon skins stretched on its outer walls, and deer and hares hanging about the door. But dreams vanish in the light of day. At breakfast on the morning following, some word indicating indecision chanced to fall, and it had the instantaneous effect of establishing the fact that there was not perfect unanimity on the subject of wintering in the mines; whereupon Mr. Murphy and myself conferred together, and in a very few moments our intention was formed of returning to our distant homes, while we had the means at command to do so. Our decision was not a great surprise to our companions, for in our discussion such a contingency had been considered. Immediately we made preparations for leaving. Our Manilla neighbors loaned us a horse to carry our baggage to Mormon Island, and Mr. Scully escorted us to that point on our journey. After crossing the South Fork, we arranged with a teamster to carry our goods to Sacramento and to deliver them at a commission store agreed upon; whereupon we proceeded on our journey.

Chapter XVIII Sojourning in Sacramento, Page 297

One day while thus engaged we chanced to meet one of the Manillanos whom as a neighbor we had known at the mines. After shaking hands, he informed us in Spanish — which we could partially understand — but more particularly by signs which he made, that one of our late companions had died. The natural conjecture of course was that the one to whom he referred was Mr. Reppert, but to be certain we at once set out for the mines.

After a toilsome march on account of the mud, and as hurriedly as possible, we reached camp on the afternoon of the following day, and found Mr. McBride sitting near the tent, from whom we learned the particulars of Mr. Reppert's death. Soon after we walked to the hill top to visit the place where our friend lay buried. A lofty oak at the head of the grave, and the broken turf covering the mound beneath which his body lay, was all that marked the spot. Mr. Murphy's experience in carving, when at sea, enabled him to prepare a suit able tablet, which was nailed to the oak; meanwhile Mr. McBride and myself brought from the river bank a number of stones with which to further mark the solitary resting place.

Messrs. Barclay and Scully had gone to Mormon Island, but returned before dusk, and were surprised, as Mr. McBride had been, to discover our return; while their greeting of course was hearty.

A marked change had come over the little group about the evening camp fire when supper was over, as we sat and smoked our pipes. All were alike sad and quiet, there was no longer any merriment, the hills had ceased to echo the loud peals of laughter. If for a time any conversation sprung up, suddenly all would relax' into silence, and sit gazing vacantly into the fire, which seemed to act as if in sympathy with the feelings of those about it ; quietly blazing up at times, *then darkening into doleful, uncertain glimmerings. Of what we were thinking, and what the intent of all, scarce needed words to reveal. That this would be the last night in which we would be engaged in camp life together was doubtless among the thoughts weighing heavily upon each mind. A unanimous conclusion, almost without discussion, was reached, that we should all go to San Francisco, and upon arriving there, those not having settled upon going home could then determine as to their future course.*

I believe the site of Johnston's mining camp was at, or very near, Condemned Bar. I would like to think the remains of Johnston's travel companion and fellow miner, George Reppert, who died in camp on the North Fork, was moved to the Mormon Island Cemetery.

Johnston would return to the North Fork Mining Camp many decades later. He recounts that much of California had radically changed, the walls of Sutter's Fort had collapsed, and all that was left was an adobe house. But the old mining camp of '49 seemed the same to his ageing eyes and rekindled fond memories of the hard work coupled with leisurely hours spent around the camp fire. He writes, "Of course I visited the site of our mining camp on the North Fork of the

American River, and there was a certain melancholy satisfaction in this. Perhaps of all places in California that I once knew, this less than any had undergone change, and I may predict with some degree of certainty that it will never be changed. Rocks and sand have so full possession that they cannot be dispossessed, and on this account I had great difficulty in going about."

No one could have predicted in 1891 that a giant dam would be erected at the confluence of the North and South Forks of the American River. But here it is, and we are left sifting through a dry lakebed looking for fragments of history and Reppert's grave site. After I read Johnston's account of the early placer mining camp on the North Fork, I wondered if I could locate its position.

Johnston states that after crossing the South Fork at Mormon's Island, it was approximately three miles to the camp, which was accomplished in one hour. He notes that the path was relatively level until they came to valley of the North Fork. From my assessment, I've concluded that his North Fork mining camp was at the base of a 540-foot-tall foothill peak at the north end of the Peninsula Campground, next to the northern Peninsula boat ramp. The tall peak is across from Dotons Point and would look like the beginning of a river canyon or valley. The road leading from Mormon Island, also fairly straight and level, would lead you almost directly to this possible mining camp. It took me about an hour to hike from the tip of the Peninsula to the base of the peak. From all accounts and old maps, Johnson's early mining camp would garner the name of Condemned Bar.

Even with Folsom Lake at historically low water levels, the shoreline was still 80 feet above the river bottom in 2015. While I knew I would never see any probable mining camp location, I thought I might be able to scout around for the site of George Reppert's grave — until I realized that so much mining had occurred in the region since 1849 that the poor guy's bones had probably been scattered long ago. I would like to think that George's remains were moved to or became a part of the Carrolton or Condemned Bar graveyards. Several grave sites and cemeteries within the outline of Folsom Lake were moved to Folsom in 1954. While the majority of the graves came from Mormon Island cemetery, the stone markers note several Unknown people who were relocated from around Folsom Lake. A listing of known deceased individuals moved to the relocated Mormon Island cemetery does not include George Reppert.

Peninsula Dams

As I hiked south from the Peninsula Campground, I was surprised at how many earthen dams there were. Some of these dams look as if they were meant to collect seasonal rains from ephemeral creeks coming out of the hills to the east. The small ponds might have been used for watering cattle or washing dirt for placer gold. Some of the dams may also have received water from the Negro Hill Ditch. I was fortunate enough to come across another set of Native American grinding holes across from Granite Beach State Park.

69. One of the several earthen dams on the Peninsula.

70. Earthen dam across from Granite State Park on the Peninsula. The dams all have familiar cuts or breeches so as to not trap water and fish when Folsom Reservoir was drawn down.

71. One of several boats that saw daylight during the 2015 drought after sinking in Folsom Lake years earlier.

Massachusetts Bar

Massachusetts Flat was above Massachusetts Bar on the North Fork and lay across from Rattlesnake Point. Massachusetts flat was the site of a placer gold mining operation. There are the concrete foundations of numerous buildings at the site. One concrete structure with walls at a 45-degree angle may have been some sort of rock hopper or sorter. The drought level of the lake revealed acres of river cobble tailings. There was also a half-submerged ore cart or wagon. Unfortunately, the silt and mud were so deep that I had no way to drag the wagon to higher ground.

72. Concrete foundation around the mining operations on Massachusetts Flat. Looking

north you can see Mooney Ridge in the background.

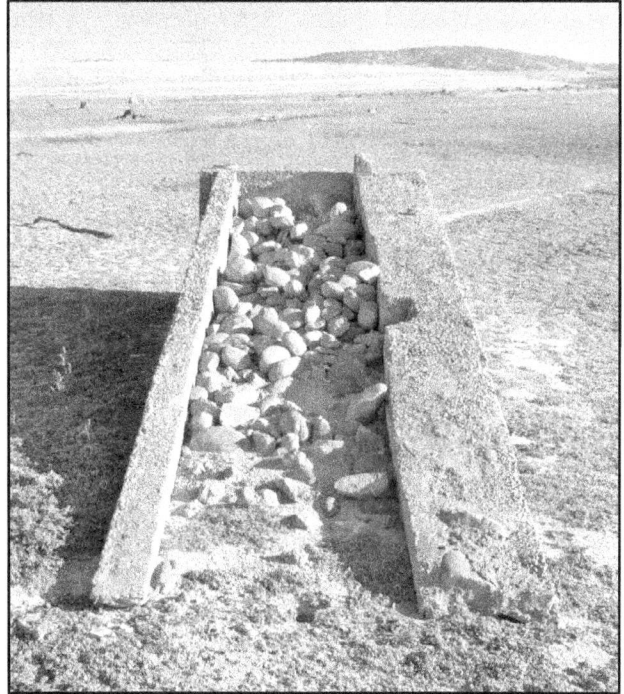

73. Elevated and angled concrete structure, possibly for separating of rock in the mining operation at Massachusetts Flat.

74. What appears to be a rock or ore cart partially submerged in the low water level of Folsom Lake among the mine tailings of

Massachusetts Flat. The lake elevation was approximately 353 feet when this picture was taken in November 2015.

75. Looking northeast toward Mooney Ridge, mounds of river cobble mine tailings cover acres of area at Massachusetts Flat.

Sacramento Daily Union, Volume 19, Number 2910, 25 July 1860

El Dorado County Correspondence

Massachusetts Flat, El Dorado Co., July 21, 1860. Messrs. Editors: Business is very dull at this place. On the 25th of February, the miners suspended work, as they say, on account of the scarcity of water. Since that time nothing has been done in the shape of mining. A correspondence passed between the miners and the ditch company, but no compromise was made. Shortly after, a proposition was made to the North Fork Company to bring water from their ditch, across the North Fork, about six miles above the place. That company refused to do it on their own account, but offered to sell the miners three hundred inches of water, and run the same twenty-four hours at twenty cents per inch. This arrangement was acceded to, and a company was formed, composed entirely of miners, who commenced digging a ditch, and the same is nearly completed to this place. The water will be brought across the North Fork on a

suspension bridge, three hundred and sixty-six feet long between the towers, and about two hundred and fifty feet above the water. When completed, the bridge will be a specimen piece of workmanship. It is being constructed under the superintendence of Calvin P. Hubbard, the architect of Lyons' bridge, over the same stream, at Condemned Bar. The water will probably be here in about two months, when it is hoped the miners will again be in successful operation, and money be more plenty. Water strikes are a great injury to any community. They not only injure the ditch company, but are a great detriment to the miners themselves. The injury to this place on account of the stoppage during the last five months will exceed ten thousand dollars.

The ditch owners in times past have been too stubborn, and in many places it has proved detrimental to their interest. It should be recollected that mining now-a-days is a hard business. In 1849 and 1850, when men used nothing but a rocker, but little water was required. But now the surface diggings are gone, and there is nothing left but deep bank claims, which require a great deal of labor and expense to prepare them to be worked. I believe in the doctrine, that the lower the fare the more will be the travel, and the lower the price of water the greater will be the demand. The average yield to the miners is not over two dollars per day to the man after paying for water and other expenses. There are some law suits pending in relation to the right of the Ditch Company to dig through the ranches.

76. Rock retaining wall of a water ditch north of Massachusetts Flat. It is unknown if this was an extension of the Negro Hill Ditch with the water flowing north toward Condemned Bar, or south from a connection to the North Fork Ditch near Carrolton, across the North Fork of the American River.

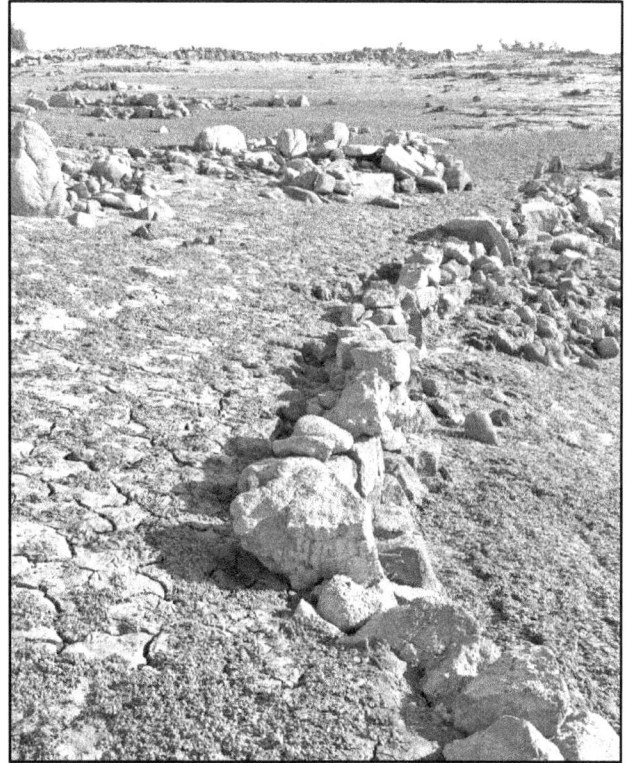

77. Outlines of a ditch that carried water to mines and farms along the eastern side of the North Fork of the American River on the Peninsula. The ditch, filled with sediment from Folsom Lake, can be traced for over a mile but doesn't show up on any maps that I could locate. This picture is looking south down the Peninsula.

The 1910 American River and Natoma Water and Mining Companies water canal map shows a bridge across the North Fork at Doten's Bar to Long Bar, a bridge connecting Carrolton to Condemned Bar, and another further south from Beal's Bar to Massachusetts Bar. There is no indication of a water pipe from the North Fork Ditch that could have delivered water into a water canal that would have serviced Massachusetts Flat. But I did come across remnants of a water ditch north of Massachusetts Flats. Whether this ditch was an extension of the Negro Hill Ditch, over which the miners were having a disagreement about water rates, or a ditch to deliver North Fork Ditch water, we may never know.

At this place on the Peninsula the term Massachusetts Flats takes on significant meaning. Along the eastern side of the North Fork the topography is anything but flat. There are numerous wrinkles in the hillside that require hiking twice the distance to get around little coves of water. This is when I began to appreciate the road from Mormon Island that travelled to the top of the Peninsula and provided a relatively straight and level hike for Johnston and his mining compatriots.

Peninsula Gold Mine

Most of the hills that have spent years submerged under Folsom Lake have a landscape of barren sand, rock, and tree stumps that have not rotted away after 60 years submerged underwater. So a hole in the side of a hill stands out to the hiker. As I approached this one small cave, a burrowing owl scampered to the top and flew away. The chain link fencing had been pulled away from what I assume was an old gold mine shaft. On top of the little bluff with the chain link fencing intact was a pit into the earth. I assume the pit and cave were connected.

78. Open pit to a gold mine is covered with chain link fencing. These holes were often referred to as coyote holes, dug by miners working to reach bedrock in search of placer gold.

Pits or shafts into the river banks and hills above the river were known as coyote holes. Miners would dig through the dirt and cobble rocks to reach the underlying granite bedrock in search of gold. There were numerous tunnels scattered throughout the area, which, as one would assume, were often dangerous for the miner.

79. The exit, or entrance, to a mine shaft that is probably associated with the open pit less than 20 yards to the south. Partially filled with sediment from Folsom Lake, it was home to a family of burrowing owls when I came upon it.

Daily Alta California, Volume 6, Number 178, 19 July 1855

Sudden Death of a Miner. — By a telegraph dispatch to the Sacramento Union, we learn that on the 17th inst. on Growler's Flat, near Negro Hill, an Englishman, named James Bennett, was killed almost instantly by the caving of the bank of his claim. His partner who was standing near narrowly escaped injury.

This Peninsula coyote gold mine looks similar in design to the pit on Gold Mine Ridge. With the prevalence of such mining activity, oftentimes obscured by vegetation or collapses, one can understand how the Bureau of Reclamation would be concerned about these subterranean mines and their proximity to the dykes and dams of Folsom Lake.

Peninsula Tip

My goal on a late November afternoon was to reach the tip of the Peninsula. I wanted to get as close to the confluence of the North and South of the American River as possible. The tip of the Peninsula resembled the description of one traveler through the area in 1860: "The whole surface of the earth has been turned over and over again, in search of gold, and now lays a scarred and unsightly monument of human energy and the love of lucre. The acres of boulders would require years in removal, and are sufficient to cobble seven cities." From the latitude and longitude of the pictures I took on the last bit of Peninsula dry land, I was a little over one mile from Folsom dam at 350-foot elevation.

80. Reaching the southernmost tip of the Peninsula, about a mile from Folsom Dam, was not the awe-inspiring moment I had hoped for. The landscape looked like any other mining operation, with piles of river cobble and tree stumps.

Peninsula Road

After miners crossed the South Fork of the American River at Mormon Island, either by skiff or later across Shaw's suspension bridge, they could travel a fairly level road across the Peninsula. The road was sometimes referred to the Georgetown Road. My other objective was to hike the Peninsula Road from as close to Mormon Island up to where I believed Johnston had his mining camp. I found where the Peninsula Road gently slipped into the lake water. I was standing on top of the bluff that overlooked the South Fork and the Mormon Island crossing, albeit submerged under lake water. It was a relatively easy hike back to the Peninsula Campground. The road is situated to run along the crest of the peninsula of land that separates the South Fork from the North Fork of the American River. I had little doubt that I was hiking close to the same path that Johnston made 160 years earlier to reach his North Fork mining camp.

81. The paved road that led to the suspension bridge at Mormon Island across the South Fork of the American River was gently submerged in Folsom Lake in November 2015.

Since Folsom Lake has smothered virtually all the vegetation beneath its waters, we must rely on the descriptions of travelers who did write about the landscape in the local newspapers. In

1860 a newspaper correspondent gave the following account of his party's travels from the Alabaster Cave down the Peninsula to Folsom.

Daily Alta California, Volume 12, Number 147, 27 May 1860

Leaving the Cave, we drove a long mile over a new road, and alongside and across limpid brooks, (affording a pleasant contrast to the muddy streams of the morning,) through a pretty green canon, to Atkins' tavern. After wine and water, we started on the Georgetown road, over the hills for Folsom. Shortly we were on the divide [Peninsula], and here a magnificent scene was presented. On the right were the Marysville Buttes; below as the valley of the Sacramento, like a map, spread out, with the river as a band of silver burnished in the sun-light; and further out in the distance, like a giant among hills, loomed old Monte Diablo. On the left were the proud Sierras, their tops covered with snow, and their crests lilting to the clouds.

Passing on the road was lined with beautiful flowers and shrubs— the chemisal, with its modest white blossom, the manzanita, the pretentious buckeye, in full bloom, the painted cup, and myriads of smaller plants. Directly after, we were in sight of Negro Hill, and in and through the town. At this place we formed a new idea of the wonders of men's work, and of what patient endurance can accomplish. The whole surface of the earth has been turned over and over again, in search of gold, and now lies a scarred and unsightly monument of human energy and the love of lucre. The acres of boulders would require years in removal, and are sufficient to cobble seven cities. Riding down the sharp hill through a deep-cut gorge, we crossed the south fork of the American River on the suspension bridge, a model of grace and beauty, and from it, swinging in space, admired a splendid landscape, embracing the hills. Mormon Inland, the river madly rushing beneath us and several ravines in the distance. A further drive of three miles, over a level plain, and we were in Folsom again. Here we lunched, started in the car at five, and were at Sacramento at six o'clock. The distance to the Cave, from Folsom, is about thirteen miles, and about eleven back, by the route we took. Next day, we returned by the steamer Eclipse, having been absent two days and five hours from San Francisco.

As I have hiked down the Peninsula several times, I concur with the author's sentiment that the views afforded on the divide are a magnificent scene. The panorama opens up as the landscape gently falls away on your left and right sides down to the rivers below. While I can't recall actually seeing the Sutter Buttes from the Peninsula, I have certainly seen Mount Diablo on a clear day.

83. Along the Peninsula road I passed by another sunken fishing boat.

84. A small concrete bridge over a drainage culvert on the Peninsula road. The road would take the traveler either up to Cool, or he/she could turn left at Condemned Bar and head up to the Zantgraf mine and the Alabaster Cave.

Negro Hill

Along the road there were some concrete structures near the shoreline of the lake. Up on

82. Square concrete-lined shaft not far from the Peninsula road. What was it used for?

the crest of the Peninsula, right off the road, were a series of concrete foundations and small diameter (approximately 1-inch) water pipes that trailed off to the east. The 1944 Folsom USGS topographical map indicates that there was a building located along the Peninsula Road near this approximate location.

85. Looking southeast toward the South Fork of the American River, on top of Negro Hill, the concrete foundation outlines of an old building. Was this the Negro Hill School House?

Sacramento Daily Union, Volume 10, Number 1477, 19 December 1855

NEGRO HILL. — A correspondent of the Placerville American furnishes some flattering evidences of the richness of the diggings on Negro Hill: Since his last letter there has been fifteen thousand three hundred and fifty dollars paid for mining claims, of which six thousand three hundred have been paid to the Know Nothing Company, for three-fourths of their claim. Since the first of this month, 367 ounces of gold dust been taken out of the hill. Money appears to be plentiful, and as the rains have increased the water in the ravines, the miners expect to reap a rich harvest during the present winter. Prospecting is being carried on pretty

extensively. Several new shafts are being sunk. The one near our office has been sunk seventy-six feet. They have a good prospect for gold. There are nineteen families residing on the hill. There are three boarding houses, one provision store, two drinking establishments, one blacksmith shop and one livery stable. There is one school house or church, in which a day school is being taught, and in which Rev. Mr. Herlbert preaches once a week, provided it does not rain or snow too hard for him to come out.

I wondered if the "school house or church" in the story could be the old Negro Hill School. However, the topographical map shows the school to be further north than the GPS location of my pictures. The 1952 aerial map of the area clearly indicates some sort of structure at the location of my pictures, but the resolution is not good enough to determine if it was a house, school or barn.

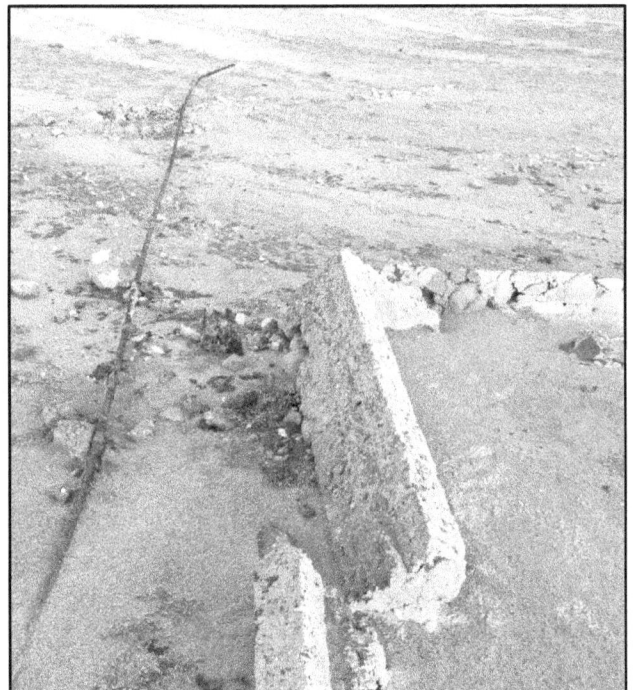

86. A 1-inch diameter water pipe runs behind the remnants of the old building on Negro Hill.

Even as temperance movements and vigilante committees formed throughout the mining camps on the North and South forks, there was still plenty of violence and thievery. As the name implies, Negro Hill was home to many African-American miners in the 1850s. While I have no doubt that racism towards African-Americans and the Chinese was rampant, there was also a segment of the white mining community that had little problem associating with black miners – and rising to their defense. I came across one story that illustrates the social dynamics between the races during the gold rush era.

Sacramento Daily Union, Volume 8, Number 1238, 13 March 1855

The Negro Hill Affray.

Negro Hill, March 9th, 1855 Messrs. Editors :— 1 saw in the Union of Wednesday, a telegraphic account of our affray here, which was incomplete. The full facts are as follows: Monday evening last, there were present in a house or drinking saloon, kept by a negro named Jackson, four whites and three or four negroes, when a gang of rowdies came in drunk and noisy; after some words, one of them seized a bench, which was pulled away from him by one of his own party; in a moment he again seized and threw it at some negroes who were standing behind a table. At the same time, a negro by the name of Henry Bell, was stabbed between the sixth and seventh ribs, by one of the rowdies. Within a minute after three shots were fired, one within the building and two without, by the same, or some other person, standing in the street and shooting through the door — two of three took effect on the person of a man named Harris, commonly known as Jimmy-from-Town, one striking above the right hip and lodging in the abdomen, the other striking and lodging in the

left shoulder. Harris is reported out of danger. Bell died on Tuesday afternoon. An inquest was called on his body, by Justice Chamberlain, and the jury returned a verdict that he came to his death by a knife in the hands of a person called Tennessee — his right name is supposed to be John Murch. The names of the perpetrators of this wanton and unprovoked outrage, besides Harris and Tennessee, are Solomon Rathboun, Moses Drew and James Comstock. The evidence elicited did not implicate Comstock and Harris as taking an active part. So far no arrests have been made, and the attention of the authorities is respectfully called to these facts. Respectfully yours, Newton C. Miller.

In March the Governor of California issued a reward for the capture of the murderers.

Daily Alta California, Volume 6, Number 98, 16 April 1855

ONE THOUSAND DOLLARS REWARD. Whereas, ON THE EVENING OF THE 5th of March, an affray occurred at Negro Hill, in the County of El Dorado, in which one Harris was wounded, and one Henry Bell was murdered; and, whereas, said murder was believed to have been committed by three men, viz: Solomon Rathburn, Moses Drew and —— Tenissen, alias John Marsh; and, whereas, the said Rathburn, Drew and Tenissen, immediately after having committed said act, fled from justice. Now, therefore, in pursuance of an act entitled "An Act authorizing the Gov. to offer rewards for the apprehension of criminals," passed April 29, l851, I hereby offer the sum of One Thousand Dollars ($1,000) as a reward for the apprehension and commitment of said fugitives from justice in the Jail of any County within the State of California. Witness my hand and the Great Seal of the State of California, at the City of Sacramento, this the 28th day of March, A. D., 1855, JOHN BIGLER, Governor.

While two of the men were apprehended in Weaverville, only Moses Drew was brought back to the area to face justice. It was later reported that Drew and his accomplices were tried and acquitted of the murder charges in *Historical souvenir of El Dorado County, California El Dorado County California with Illustrations and Biographical Sketches of its Prominent & Pioneers*, by Paolo Sioli. Oakland, Cal. 1883, Paolo Sioli Publisher

Sacramento Daily Union, Volume 9, Number 1284, 7 May 1855

Murder. — A man named Moses Drew came down on the Helen Hensley [steam ship] yesterday in charge of Sheriff Buel, of El Dorado County, and was securely lodged in the station house for the night, to be conveyed mountainward today. He was arrested at Weaverville, Shasta County, bound for Oregon, on the 19th April, and is charged and indicted as being one of the whites who were engaged in the affray at a negro house at Negro Hill, (near Mormon Island) on the 5th of March last, resulting in the death of a negro named Henry Bell. The difficulty in question is said to have been provoked by the defendant and several other whites, who entered the house in a drunken and riotous manner, and made a brutal assault on the inmates. A reward of $1,000 was offered by the Governor for the arrest of the murderers, or either of them. The others have not been taken.

I could not find a grave at the relocated Mormon Island Cemetery that bore the name of Henry Bell. He may have been in one of the graves from Negro Hill relocated and marked Unknown.

In 2011, new grave markers with the appropriate designation of Negro Hill replaced the "Nigger Hill" grave markers placed by the U.S. Government in 1954 at the Mormon Island relocation cemetery.

Negro Hill Ditch

Remnants of the Negro Hill Ditch are hard to spot on the Peninsula. This water canal begins up by Salmon Falls on the South Fork. It didn't supply nearly as much water as its counterparts the Natomas Ditch or the North Fork Ditch. It isn't even illustrated on the 1910 Map of the American River and Natoma Water and Mining Company's Canals, indicating that it might have been owned by another entity. However, later USGS topographical maps of the region do illustrate it and its course around the Peninsula. I did come across the outline of the ditch in the dry lakebed, which closely mirrors its location on old USGS topographical maps. In addition, I located some of the rock retaining walls used on the water canal.

Sacramento Daily Union, Volume 7, Number 1014, 23 June 1854

The Diggings— Mining Ditch—Post Office, &c.

Salmon Falls, June 21, 1854. Messrs. Editors: — Permit me through your column, by way of novelty, to furnish you and the public in general

with a few of the incidents that at present occur in and about this old place of '49.

The miners in and about Salmon Falls are averaging very small wages, with the exception of Higgins Point, situated about a mile below the Falls, where they are making from $5 to $6 per man, daily, and have for a considerable time been doing well; but since the excitement has occurred as regards the diggings on the south side of the North Fork, opposite Condemned Bar, we are becoming deserted. I was there yesterday, and persons are flocking in from all parts, waiting with patience and anxiety until the advent of the water from Boyd's Bar Water Company. This company are lengthening their ditch, which has for the past year supplied water on the old bar opposite Salmon Falls.

For extent and general average of diggings, in my humble opinion, there are few places to cepe [compete] with them in the mines. There is ground enough to last for years for any number of persons, having been thoroughly prospected, there being no doubt that they will average excellent wages. When the water arrives and the claims are in full operation, if you should ever visit the mines and go there, you will see as flourishing a place as there is in the mountains. The cost of cutting said ditch will not fall short of $30,000.

If this article is correct, a company by the name of Boyd's Bar Water Company was extending the water ditch that started above Salmon Falls. However, the author places Condemned Bar on the north or west side of the North Fork. It's unclear if the Boyd's Bar Water Company ditch ever extended the canal as far north as Condemned Bar opposite Carrolton. All maps only indicate one water ditch extending onto the Peninsula. The Negro Hill Ditch began at

approximately 440-foot elevation, along with the Natomas Ditch, which used the same dam on the South Fork and is shown terminating at 375-foot elevation on the 1941 USGS Folsom topographical map.

Mormon Island to Salmon Falls

The town site of Mormon Island lies at approximately 260 feet in elevation, so there was little hope that the lake would ever drop low enough to expose any of those buildings. The Bureau of Reclamation states that the top of the inactive Folsom reservoir is at 327-foot elevation. Near where the Bureau of Reclamation was adding a dam spillway on the eastern side I did stumble across what looked like an old pump house with engine. There was a water canal called the Mormon Island Ditch that was a branch from the main Natomas Water Ditch. This pump house may have been associated with that water infrastructure.

87. A small rock wall dam may have stored water released for the Natomas Ditch up the hill at Mormon Island.

Forever submerged beneath Folsom Lake are the remnants of a wire suspension bridge built in 1855 to connect Mormon Island to Negro Hill over the South Fork of the American River.

Daily Alta California, Volume 6, Number 181, 23 July 1855

LETTER FROM MORMON ISLAND.

Mormon Island, July 20, 1855.

In compliance with your request I pen you a few lines for the purpose of keeping you posted in this vicinity, although nothing of much importance has transpired of late. Business is rather dull at present. Petty robberies are quite common among the miners. Chinamen have suffered considerably from the thieving rascals. In one fracas a Chinaman was killed instantly for striking a light while the rascals were plundering the cabin. Miners are doing pretty well. River mining has just commenced and looks very favorable. Politics on the ebb. The temperance cause is gaining ground very rapidly: a new division was organized at Negro Hill last Sunday evening.

J. W. Shaw, Esq., of this place has nearly completed a suspension bridge over the South Fork of the American River which, when finished, will surpass anything of the kind in this section of the country. I will give you the details of the same when I obtain them of the Engineer, Mr. Bronk.

Very respectfully I remain yours, & c., Leeman Smith, Agent Wells, Fargo & Co.

By 1857 the once bustling Mormon Island had calmed down. While gold mining was still occurring, the methods had changed, and the focus had turned from the river bars to placer benches in the hills above the river.

Daily Alta California, Number 54, 24 February 1857

Owing to the destruction of Kinzie's bridge, which spanned the American at this place [Folsom], the Placer and Nevada stages, and, in fact, all travel through those counties, are compelled to cross the suspension bridge at Mormon Island, four miles above Folsom, on the South Fork of the river. This latter bridge is a graceful and beautiful structure, built entirely of iron, on a foundation of heavy stone abutments, and elevated far above high water mark. It is the property of a Mr. Shaw, and was erected at a cost of ten thousand dollars. The proprietor, in consequence of the disaster which has befallen his enterprising neighbors, is reaping a rich harvest; but the latter have already made arrangements for the construction of as elegant and costly a suspension bridge at this point as is the one just referred to.

Mormon Inland exists but in name, for almost every vestige of the island has disappeared before the pick and shovel of the miner, and the waters of the Fork, which have cut eccentric channels through the original land from which the place in part derives its name. The settlement is dull, and on the street which five years ago presented not a few stores, hotels, restaurants and saloons, there are to be seen now but scattered cabins and dwellings of frame, beyond the fireproof Pacific Express building, Wells, Fargo & Co.'s office, and the very excellent hotel, which is fully equal to any in that section of the country. The saloon of the town, owned by a darkey, rejoices in the significant title of "Forty Drops." Whether that is the maximum of the amount of spirits permitted to be imbibed for a solitary "quarter," or whether this quantity is all that is required to " kill at forty yards," deponent saith not.

Mormon Island Cemetery

The old Mormon Island Cemetery sat on a hill overlooking the town just north of the Dyke 8 picnic area at about 370-foot elevation. Even though all the remains were relocated in 1954 to the south side of the dam, the old family plot concrete outlines still remain. All the Mormon Island graves along with cemetery residents from Salmon Falls, Negro Hill, Dotons Bar, Carrolton, and McDowell's Hill that would have been underwater when Folsom Reservoir filled were moved to Shadow Fax Lane in the City of Folsom.

88. One of the many family burial plots outlined with cement and stones on the low hill that overlooked the town of Mormon Island, revealed by low water levels at Folsom Lake.

Mormon Island School

East of the Mormon Island Cemetery is a large foundation that maps indicate was the Mormon Island School. The foundation sits at approximately 360 feet of elevation and was situated on a county road. The foundation is north of the Mormon Island Dam and west of Brown's Ravine.

89. The concrete foundation of the Mormon Island School House with a drought-shrunken Folsom Lake in the background.

Richmond Hill

The Natomas Ditch, which had its dam above Salmon Falls, runs around Brown's Ravine at about 375 feet of elevation. The Natomas Ditch overlooks the Richmond Hill mining district and the community of Red Bank. The 1910 Water Canal Map indicates there were a couple of branch lines into Richmond Hill and Red Bank. Similar to the area around Beals Point, there was so much construction for the Mormon Island Dam and the previous placer gold dredging operation, it's hard to tell what is original 19th century construction and what was built later.

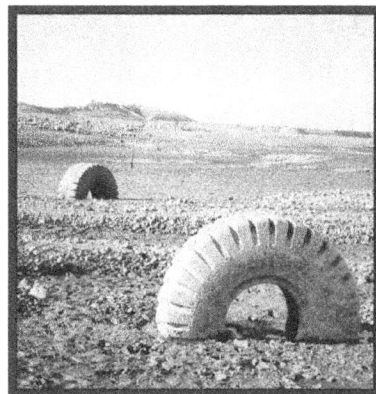

Old tractor tires left behind from the construction of Folsom Dam dot the landscape on the northern end of the Richmond Hill mining district.

Daily Alta California, Volume, Number 54, 24 February 1857

In the immediate vicinity of Mormon Island, are Richmond Hill, Negro Hill, and Beale's Bar, all of which are strictly mining villages, and together, number about three hundred diggers. Tunneling and sluicing are extensively carried on, and "underground railroads" thread every portion of this subterranean region. Although the immediate vicinity of Mormon Island was upturned, and as was supposed at the time, thoroughly worked at the very earliest period in the history of California gold digging, yet today there are companies on Richmond Hill, making on an average, one ounce per diem to the man; thus showing that the Placer diggings of this State are in fact inexhaustible. Above Negro Hill and on the peninsula formed by the North and South Forks of the American river, rich auriferous quartz has been found, and judging from the energetic character of those who have laid claim to the discovery, there is little doubt that they will in time be amply remunerated for their expenditure of time and capital.

Red Bank

The community of Red Bank blossomed after the initial gold rush. In addition to a number of homes, there were also dairy, vineyard and winemaking operations. In 1873 it was reported that 2,000 gallons of wine from the Red Bank Vineyard, near Folsom, were sent to St. Louis. The surrounding Red Bank community was active in raising money for their local schools.

Sacramento Daily Union, Volume 7, Number 67, 8 May 1878

Last Friday a May festival was given at Mormon Island in aid of the Natoma District School, and it proved to be one of the most enjoyable and best conducted affairs of the kind ever held in this neighborhood. The weather was delightful and the beautiful grounds, with their carpet of green grass and groups of happy children, were charming to behold. At noon the exercises of the school children took place and reflected very great credit on Miss Addie Davis, their teacher, for the effective way the little ones were trained. Mr. and Mrs. Henry Mettee gave a grand ball on behalf of the school at their beautiful place at Redbank Vineyard, giving with great generosity their large — in fact, their whole establishment — free of any charge whatever. The ball was well attended and proved a financial success, netting $100.

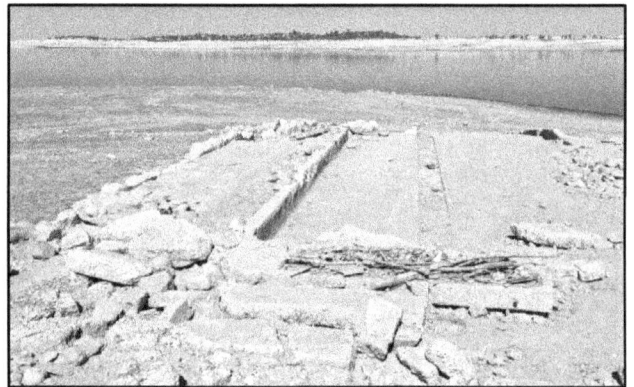

90. The remains of the old Red Bank Dairy facility, looking northeast toward Granite Bay.

91. Another view of the dairy barn showing how the local river cobble rocks were incorporated into foundations of buildings.

92. Before the advent of inexpensive concrete, home and barns had their foundations constructed of local stones carefully stacked like the remnants of this structure in Red Bank.

There are several foundations of stone rubble and what appears to be an old dairy building still visible in the Red Bank area when the lake level drops below 375-foot elevation. While this area was choked with people in 2014, most of the local area residents had lost interest in dewatered historical relics of Folsom Lake by 2015. The Natomas Ditch runs above the community of Red Bank and the Richmond Hill mining district. You can follow the ditch north to a spot midway

between the Mormon Island Dam and Brown's Ravine boat launch.

McDowell Hill

In between Red Bank and Higgin's Point was the tiny community of McDowell Hill. I only learned of it while walking through the relocated Mormon Island Cemetery. Of the resources I consulted, McDowell Hill only shows up on the 1910 Water Canal map. There was a short branch line canal from the Natomas Ditch to the McDowell's Hill Community.

93. Looking west down the South Fork of the American River from atop the Natomas Ditch with weir outlet in foreground just south of Higgins Point. It was at this point the flowing South Fork met the still waters of Folsom Lake in 2015.

Sacramento Daily Union, Volume 5, Number 658, 3 May 1853

New Diggings. — We understand from a gentleman residing near Mormon Island, that new diggings of great extent have just been discovered between that place and McDowell Hill, yielding an average of from three to five cents to the bucket. Some of the prospects have been as high as 50 and 60 cents. The canal of the Natoma Company has been completed within a few days to Mormon Island, and is now supplying that whole section with water, at which point miners are congregating from all quarters. We are also informed that the Natoma Company are extending their canal as rapidly as possible to Rhodes' and the Willow Springs, and fully expect to have the water there by the 1st day of June. This work will open one of richest and most extensive mining regions in California.

Higgins's Point

Three miles up from McDowell Hill on the South Fork is a horseshoe bend in the river around Higgin's Point. Like McDowell Hill, Higgin's Point only shows up on the 1910 Water Canal Map, which did indicate a building of some sort. Around Higgin's Point the Natomas Ditch is running at about 425-foot elevation, the Higgin's Point structure is approximately 400 feet in elevation, and the river bed is 360 feet in elevation. Old topograhical maps indicate that the horseshoe river bend was the actual Salmon Falls. In November of 2015 a free-flowing South Fork met the still waters of Folsom Lake just downstream of Higgin's Point.

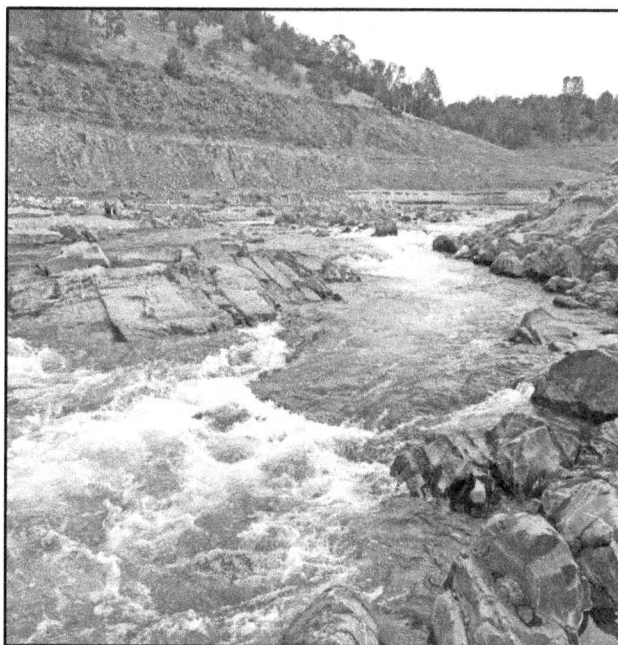

94. Looking east, at the tip of Higgin's Point the South Fork of the American splashes over bedrock that would be coined Salmon Falls. On the opposing hillside is a line that was the Negro Hill Ditch. To the south up the hill was the Natomas Ditch.

Sacramento Daily Union, Volume 7, Number 1014, 23 June 1854

The miners in and about Salmon Falls are averaging very small wages, with the exception of Higgin's Point, situated about a mile below the Falls, where they are making from $5 to $6 per man, daily, and have for a considerable time been doing well;

mounds of rocks stacked up on the north side of the river.

Natomas and Negro Hill Retaining Walls

It was something of a mystery as to why there were mounds of rocks in discrete piles on the north side of the South Fork across from Higgin's Point. Their height and arrangement indicated they were placed or stacked by the hand of man. These rock mounds were not a river phenomenon, nor did they seem to be burial mounds. The more plausible and mundane conclusion I have drawn is that these mounds were the collection of rocks used in the retaining walls for the Natomas and Negro Hill water canals.

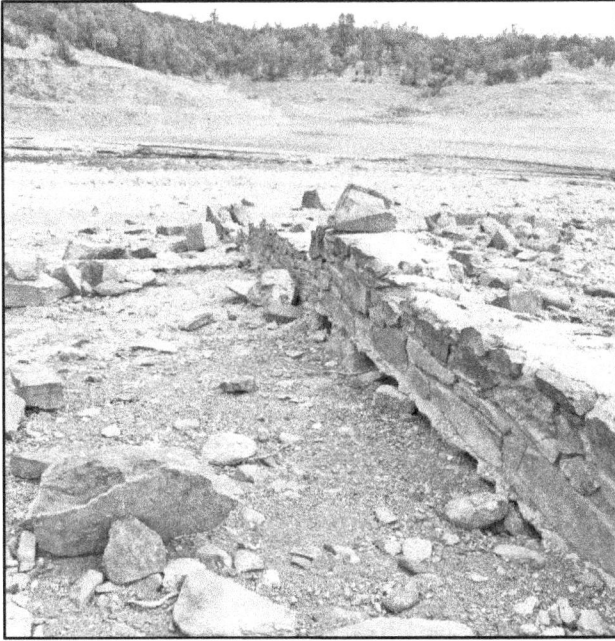

95. The stone and cement remnants of a house that sat on top of Higgin's Point overlooking Salmon Falls.

In the summer and fall of 2015, the South Fork of the American River was flowing freely around Higgin's Point. The river had pushed a lot of the silt downstream, and the bedrock jutting up from the river bottom creates little water falls. If there were any salmon to migrate up the South Fork, I could see them jumping over these river waterfalls, which may be the origin of the name Salmon Falls. The steep walls of the hills and the way the water crashes over the riverbed make this a beautiful spot to hike to when the lake is low.

The geology of El Dorado County where the South Fork runs through it is different from the North Fork in Placer County. Around Higgin's Point one rarely sees much granite but instead more metaphoric bedrock. The South Fork runs through a canyon in this area until the canyon widens upstream in the Salmon Falls community area. While I found no Native American grinding holes in the area, I did find a number of odd

96. The stacks of angular blocks of stone supported the Negro Hill Ditch on the north side of the South Fork of the American River.

When you see the height and length of some of the dry stone construction retaining walls you will understand how many rocks the builders needed. Because of the steep slopes of the river canyon, it is not uncommon to have walls 15 feet tall and 100 yards long, curving with the hillside. There is no mortar in the wall construction. It is just one carefully selected stone placed next to another — all laid by hand with no machines in the 1850s.

97. Over fifteen mounds of local rocks dot the landscape on the north side of the South Fork. It's my speculation that the rocks were gathered as a source for the stones used in the retaining walls to support both the Negro Hill Ditch and Natomas Ditch.

While the rock walls weren't necessarily engineered, common sense dictated a specific pattern of rock sizes be used from the base to the top. Thousands of rocks had to be gleaned and staged for the contractor to select the appropriate size when building the retaining walls. Hence, mounds of rocks, piled near their gathering spots, for the builders to select from.

98. Local rocks were used to support the concrete lining of the Natomas Ditch. I saw no evidence that the Negro Hill Ditch was ever lined with concrete.

When the lake level drops to 380-foot elevation, you can clearly see the rock retaining walls as they hug the hillsides and supported the old water canals. While it can be a challenging hike over the abandoned water ditch, walking on its top with the river running free below gives you a unique perspective on the difficulty of constructing the retaining walls so far above the river. The river from the old Salmon Falls Bridge to around Higgin's Point falls 50 feet in elevation within one mile. This elevation drop resulted in the water rushing over the jagged bedrock to create the Salmon Waterfalls of yester-year.

99. Looking west, north of Higgin's Point, the Natomas Ditch, 30 to 40 feet above the South Fork, hugging a steep hillside.

The beauty of the water canal next to the rough and tumble waterfalls in the river was not lost on Emilie Connelly when she wrote about her hike in 1908.

Sacramento Union, Number 187, 30 August 1908

By Emilie Connelly

Leaving Newcastle late in the afternoon, we drove along a most beautiful road, across the North Fork of the American, past Alabaster cave, which was a favorite destination for tallyho parties in years past, and which is today just as astonishing and as beautiful; past the new lime kiln, up and down hill, but constantly climbing until, just as the shades were falling, we drove down into the canyon, where the road runs into the Middle Fork of the American. It was an ideal camping place, and previous parties had left a stove of stones, so we were ready for the night, sleeping out under the stars.

The moon rose late, flooding the canyon with silvery light; the water rippled softly over the stones, foxes barked; we heard the cry of a bittern and other strange birds. Early in the morning one of the party found the tracks of a large buck nearby, and later a coon was treed and killed. We were in the wilderness, and yet only thirty-five miles from Sacramento.

After breakfast we started to Salmon Falls, enjoying one of the most beautiful walks I have ever taken. First it was necessary to ford the river, stepping from rock to rock, and after wetting our feet in the crystal water there came a bit of beautiful road under an arch of forest trees and we had reached the Irrigation ditch, beside which ran a well-beaten path. For about the length of a city block we walked over a plank above the swiftly flowing water and passed several water-wheels which carry the water into the consumers' pipes.

Finally our path lay again beside the ditch, which ran around a hillside, and resembled some quiet creek. On one side rose the wooded mountain; on the other, forty or fifty feet below, the river ran over its rocky bed. Beyond the hills rose again in graceful outlines. The little path was shaded by forest trees, and near the end we came to a well-kept farm with a bit of orchard and a deliciously cool spring.

The water in the falls is low, but they are still beautiful, and the river gorge itself, with its piles of cracked and broken slate, if pictured to us on a European post card would call forth enthusiastic admiration. For some time we sat on the rocks, watching the foaming waters as they tumbled over in three separate streams, and then reluctantly turned our faces homeward. We live in anticipation of seeing Salmon Falls sometime in the spring, when the river is a raging torrent and the hills are covered with blue bells and poppies.

From the north side of the South Fork, looking at the long rock retaining wall that supported the Natomas Ditch just south of the community of Salmon Falls upon which Ms. Connelly probably strolled.

Salmon Falls Bridge

Because the South Fork river canyon opens up and flattens out north of Higgin's Point, it was a good location for a bridge across the river. The Salmon Falls bridge that is revealed during the low lake levels was constructed in the 1930s. South of the bridge on either side of the river appear to be concrete footings for utility poles to carry wires across the river. The water ditches have been filled with sediment and silt or have fallen apart along this part of the river. On the north side of the South Fork, the county road to Georgetown and Cool can be seen.

100. This black-and white-photo looking east of the Salmon Falls Bridge looks like it might have been taken when the structure was in use in the 1950s. The bridge's condition is still very good even after being submerged beneath Folsom Lake for so many years.

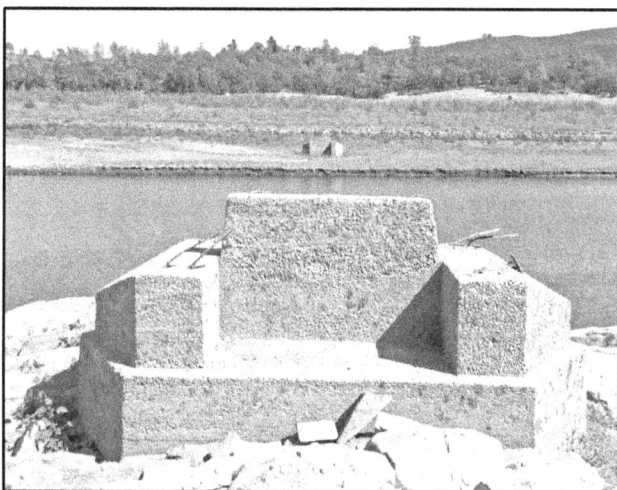

101. Large concrete foundations on either side of the South Fork at Salmon Falls were probably for utility lines crossing over the river.

Negro Hill Ditch and Aqueduct

The Natomas Ditch ran on the south side of the South Fork, and the Negro Hill Ditch ran on the north side. The Negro Hill Ditch was a small canal and only carried water as far as Negro Hill and a little north on the Peninsula. The Natoma Ditch had multiple branch lines and ultimately terminated down by Prairie City, south of the City of Folsom. Both ditches started at a dam on the South Fork, which is south of the current Salmon Falls Bridge by the State Park parking lot for river rafting takeouts.

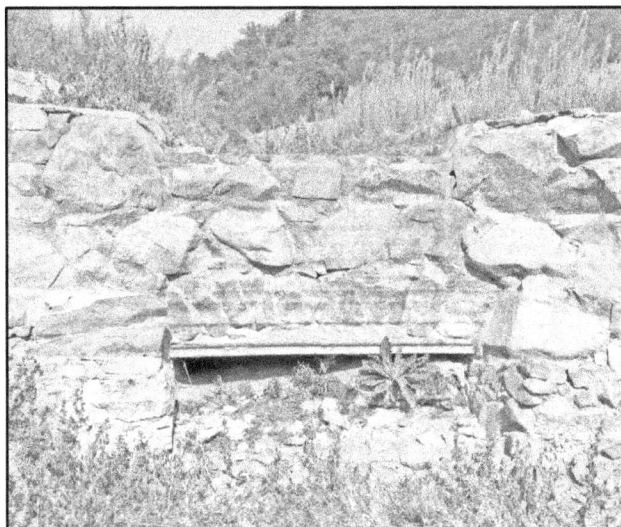

102. A short aqueduct on the Negro Hill Ditch made of local stones and cement spans Indian Springs Creek on the South Fork of the American River.

Navigating the water ditches around or across steep ravines and creeks was problematic for the early canal contractors. If you dug the ditch into the ravine, it could easily be washed out by seasonal creeks running down the crevice of the hillside. The alternative was to build an expensive flume to span the ravine. None of the flumes indicated on old maps exist today. I did stumble across one short flume that, because it was constructed of rock, has withstood the ravages of high creek flows and being submerged under Folsom Lake.

103. Supporting the weight of the rocks as the aqueduct spanned the creek are railroad track rails.

Located approximately one-half mile downstream of the water canal dam on the South Fork is a short stone rubble masonry aqueduct over Indian Springs Creek. This built-to-last flume on the Negro Hill Ditch is only 20 feet in length. What is interesting is that they used old railroad rails as horizontal supports to support the rock flume across the creek.

104. The Negro Hill Ditch stone aqueduct is less than 100 yards from the South Fork. Unlike the stone aqueduct on the North Fork

Ditch, it was sealed with concrete and actually carried water.

Natomas Ditch Canal and Dam

The South Fork dam that diverted water into the Natomas and Negro Hill Ditches was not as big as the Birdsall Dam on the North Fork. From 1952 aerial photography, the low-rise dam constructed of rock blocks is clearly visible. This is the same material that is present today when the reservoir level is low and you can walk across the remnants of the old dam. The Natomas Dam was 285 feet across the river and rose 12 feet from the riverbed. It was constructed of quarried granite and reinforced with 3/8-inch rod iron stakes and cement. When I was able to inspect the remnants of the old dam, I could see the large iron rods placed in drilled holes of the rock for additional support against the force of the flowing river.

1952 aerial photograph shows the South Fork of the American River flowing over the Natomas Dam.

Directly below the state parking lot, on the south side of the river, is a rusty paddle wheel that was a fish and debris screen for the canal. The only markings I could find on the Natomas Ditch were some initials impressed into the concrete lining west of the paddle wheel. The dam was breached with charges of dynamite as Folsom Lake began to fill.

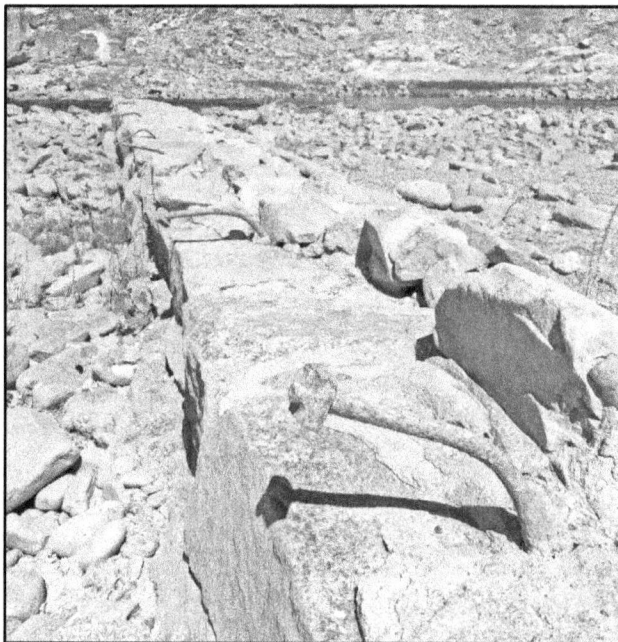

105. The stone and bolt remnants of the Natomas Dam.

It was at the paddle wheel that I forded the South Fork, like a real man. And, like a dumb old man, I slipped in the river and took a bath. However, because I understood that technology is more important than life or limb, I was able to hold my smart phone out of the water, primarily because I was recording the event.

106. Only a narrow strip of the Natomas Dam base material remains, after being blown up with the impending rise of Folsom Lake, indicating the line across the river.

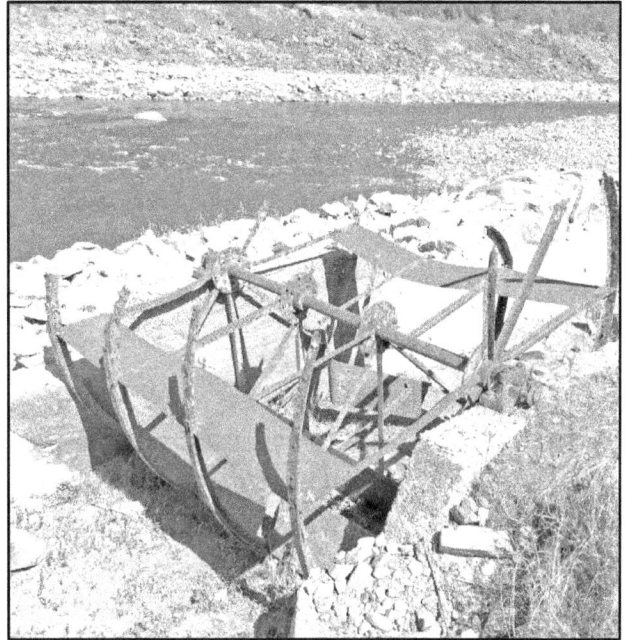

108. The rusty remnants of a paddle wheel fish and debris screen on the Natomas Ditch.

107. Concrete canal of the Natomas Ditch a quarter-mile below the dam.

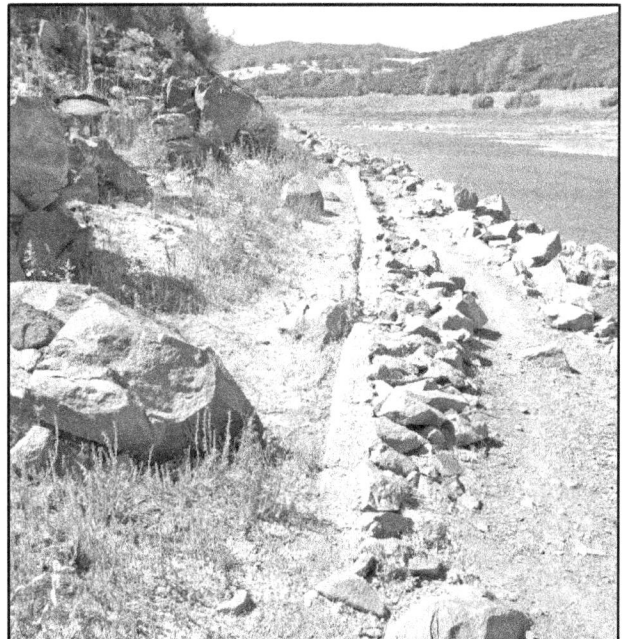

109. Natomas Ditch was lined with concrete, but the original rock retaining walls still remain. Now filled with sediment from Folsom Lake, the Natomas Ditch makes a nice level hiking path down to the Salmon Falls Bridge when the water is low enough.

Conclusion

While it was fun to stumble across history made by man over a century ago, it was also humbling to witness nature in the absence of lake water. I realized after I read Emilie Connelly's 1908 account of her hike to Salmon Falls that most people will never see this sight of nature.

The water in the falls is low, but they are still beautiful, and the river gorge itself, with its piles of cracked and broken slate, if pictured to us on a European post card would call forth enthusiastic admiration. For some time we sat on the rocks, watching the foaming waters as they tumbled over in three separate streams, and then reluctantly turned our faces homeward. We live in anticipation of seeing Salmon Falls sometime in the spring, when the river is a raging torrent and the hills are covered with blue bells and poppies.

I saw the falls from high above the river as Emilie did. I sat by the river as it tumbled over the broken slate of the river bottom. I, too, reluctantly had to turn my face homeward. But it is doubtful that I, or anyone else, will be able to return and re-experience this scene, as this part of South Fork of the American River is normally covered by Folsom Lake. But for a brief time during the drought of 2015, I was truly able to step back in time to see and hear a river running free.

As much as I reveled in watching the river rush over the Salmon Falls, I was equally saddened at the destruction to the riverbanks by gold mining as the lake water receded. Whole swaths of river bank had been torn away either by hydraulic mining or mechanical means, leaving ugly scars. The tailings of river cobble were piled up in mounds no different from areas along the lower American River below Folsom and Natomas dams. As an amateur historian, I had romanticized about finding a lake bottom and river banks similar to what the early gold miners might have seen. It is naïveté and romanticism that often separate the novice from the professional historian.

By the time Folsom Dam and reservoir were proposed, the North and South forks of the American River had been the sites of industrial placer mining. Hardly anyone lived in the area, and its economic value as a gold producing region had played out years earlier. What better place for a dam than at the confluence of two rivers that few people valued any longer.

Folsom Lake is good at smothering history. The thousands of people that boat, bike, ride, run, hike and picnic around the lake have little understanding of the history that lies beneath the surface of the water. I do not fault those who use the lake as it is presented today, with scant reference of yesteryear. Whether Folsom Lake is full or nearly empty, it offers a wide range of recreational opportunities. I would rather see people out on the lake than sitting in a room watching television.

I would like to think the historically low water levels experienced in 2015 at Folsom Lake are a once-in-a-lifetime event. Unfortunately, California continues to grow in population, as does the demand for water. Add to this the uncertainties of climate change, and I might be out hiking the dry lake bed of Folsom reservoir looking for history again in the near future.

Maps

There is no single topographical map that encompasses all of Folsom Lake from confluence of the North and South forks of the American River up to the diversion dams for the water canals. To illustrate the general areas where I hiked and discovered various historical sites, I have included reproductions of USGS topographical maps that were drawn while Folsom Dam was being built, illustrating where the high water of the lake was. Indicated below each map are the relative locations of interesting historical features and pictures included in the book.

Folsom Dam to Mooney Ridge

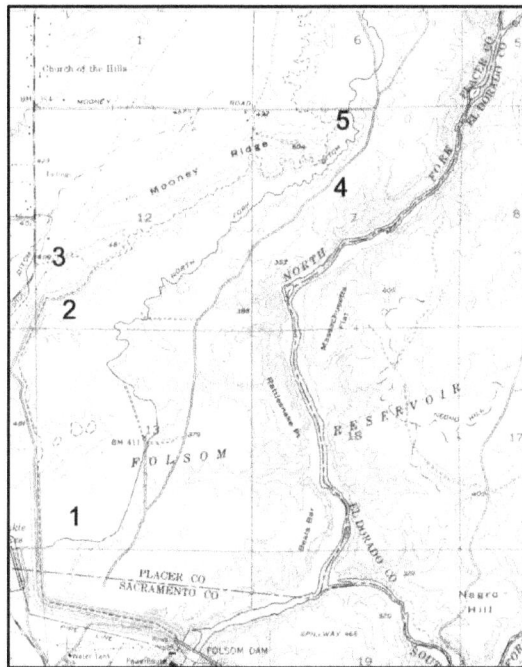

1. North Fork Ditch Reservoir and canal images
2. Rose Springs, dump site and drainage cut
3. Gold Mine Ridge
4. Native American grinding holes
5. Rock Springs

Granite Bay to the Narrows

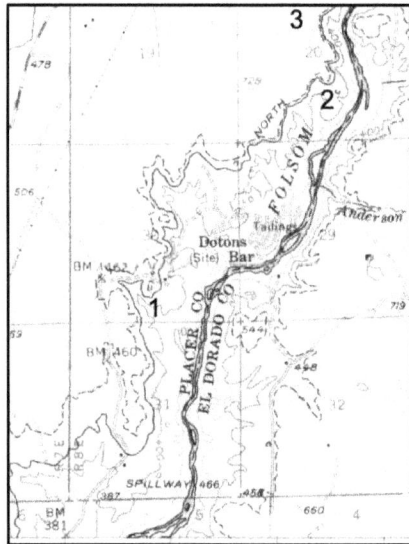

1. Beeks Bight
2. Quartz or limestone outcropping
3. Narrows, suspension bridge

Horseshoe Bar, Rattlesnake Bar, Wild Goose Flats

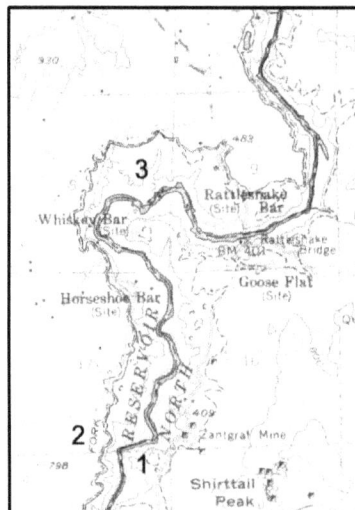

1. Anderson Island, suspension bridge
2. Stone mason aqueduct
3. Native American grinding holes

Mormon Bar to Birdsall Dam

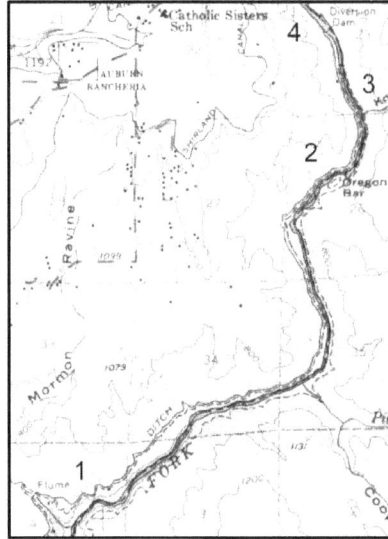

1. Mormon Bar
2. Multi-weir concrete structure
3. Knickerbocker waterfalls
4. Birdsall Dam

Peninsula, Mormon Island

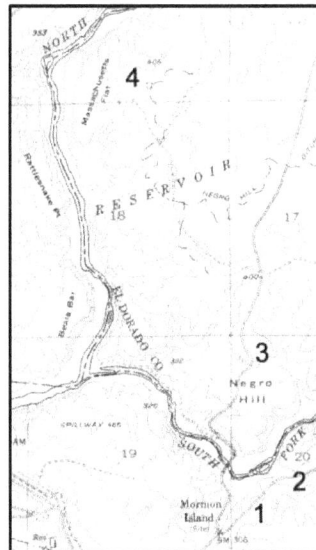

1. Mormon Island Cemetery
2. Mormon Island School
3. Negro Hill building foundation
4. Massachusetts Flat mining operation, water ditch

Richmond Hill, Red Bank

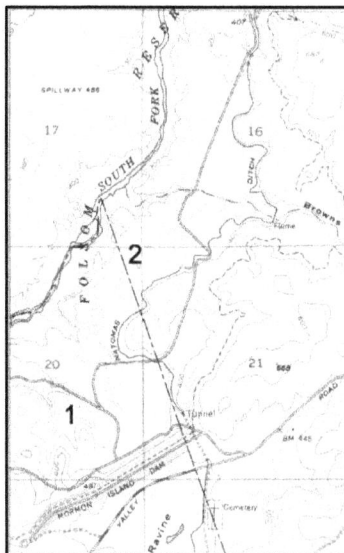

1. Richmond Hill mining district
2. Red Bank dairy and foundations

Higgin's Point, Salmon Falls, Natomas Dam

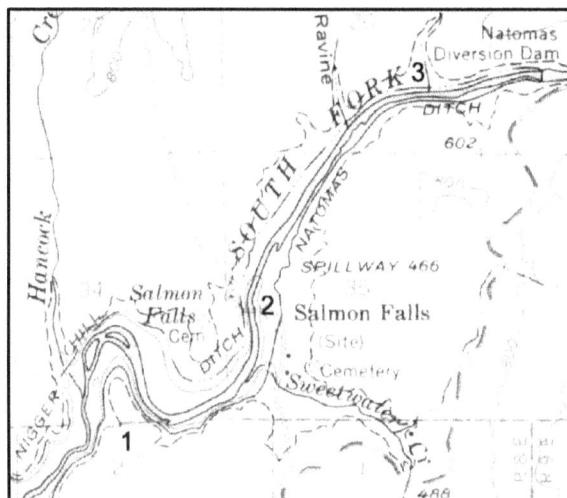

1. Higgin's Point
2. Salmon Falls Bridge
3. Negro Hill stone aqueduct

1909 North Fork Ditch

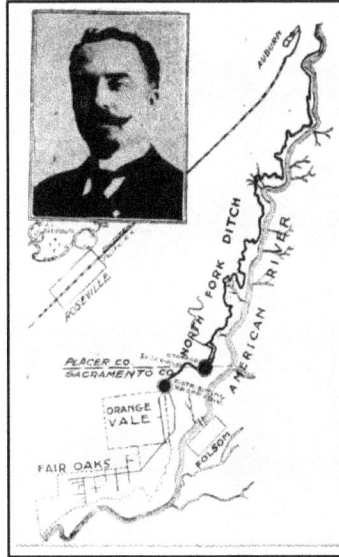

Pictured is A. L. Darrow, president of the Fort Sutter National Bank, who was negotiating for the purchase of the North Fork Ditch. Sacramento Union, March 9, 1909.

1910 American River & Natoma Water & Mining Company Canals

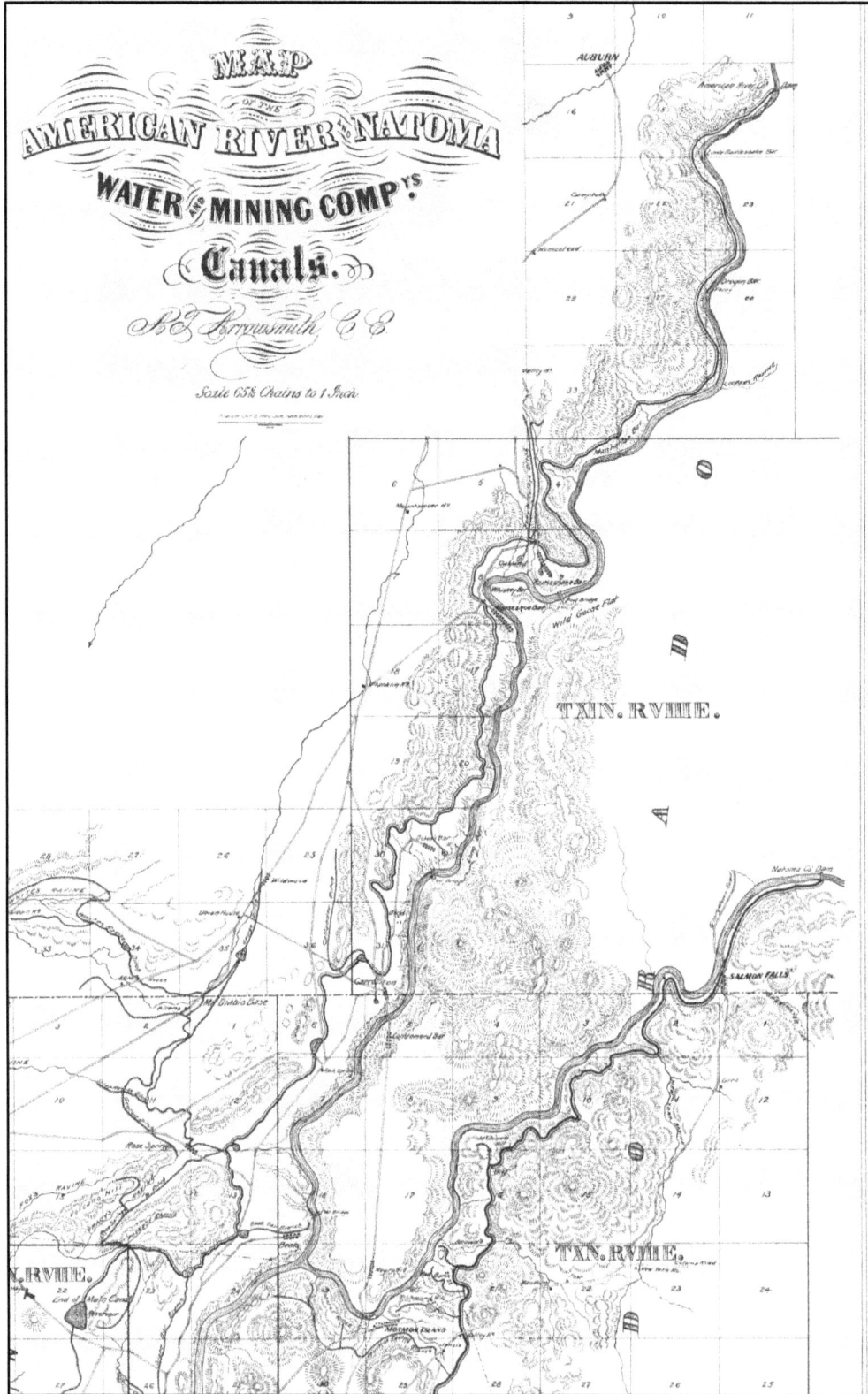

Picture Locations

Approximate GPS locations of pictures included in this book. The GPS locations were captured by my camera phone and may not always be reliable. Many of these photos are posted to my Panaramio account, InsureMeKevin, and may be viewed through Google Earth.

Picture Number	Description	Degrees; Minutes; Seconds	
		Latitude North	Longitude West
1	Folsom Dam low water	38; 42; 38.47	121; 9; 28.84
2	Folsom Dam spillway	38; 42; 12.28	121; 9; 36.95
3	Beals Bar branch ditch	38; 42; 55.88	121; 9; 37.63
4	Beals Point NFD outlet	38; 43; 2.79	121; 9; 43.87
5, 6, 7	North Fork Reservoir distribution	38; 42; 53.73	121; 10; 1.6
8	Bridge timbers by Dyke 5	38; 43; 41.40	121; 10; 2.4
9	North Fork Ditch, Beals Point	38; 43; 2.4	121; 9; 44.4
10	1915 concrete date	38; 43; 29.76	121; 9; 39.8
11	Flume support	38; 43; 18.83	121; 9; 36.75
12	Top of Dyke 5	38; 43; 47.39	121; 10; 12.59
13	Dyke 5 drainage cut	38; 41; 57.6	121; 6; 43.2
14	Rose Springs	38; 43; 46.86	121; 10; 7.38
15	Rose Springs pond	38; 43; 42.35	121; 10; 3.81
16, 17	Gold Mine Ridge pit	38; 43; 54.6	121; 10; 6.6
18	North Fork Ditch concrete lip	38; 43; 49.76	121; 9; 39.47
19	North Fork Ditch service	38; 43; 46.57	121; 9; 38.57
20	Concrete farm foundation	38; 43; 42.89	121; 9; 20.28
21	North Fork Ditch rock support	38; 43; 45.97	121; 9; 40.29
22	North Fork Ditch granite cut	38; 44; 28.37	121; 8; 34.71
23	Native American grinding holes, Mooney Ridge	38; 43; 58.53	121; 9; 1.59
24	Mine tailings Mooney Ridge	38; 43; 32.73	121; 9; 2.09
25	Rock Springs	38; 44; 34.53	121; 8; 33.5
26	Rock Springs house foundation	38; 44; 26.85	121; 8; 30.45
27	Carrolton mine tailings	38; 45; 57.27	121; 7; 26.13
28	Carrolton North Fork Ditch weir	38; 45; 31.89	121; 7; 38.35
29	Beeks Bight concrete foundation	38; 46; 14.98	121; 7; 32.75
30	North Fork Ditch Dotons Bar	38; 47; 32.99	121; 6; 31.79
31	North Fork Ditch Doton's Bar weir outlet	38; 47; 34.01	121; 6; 30.36
32	Dotons Bar mine tailings	38; 46; 30.99	121; 7; 13.74
33	Dotons Bar pillar	38; 46; 34.98	121; 6; 50.72
34	Cast-iron riveted pipe	38; 48; 2.28	121; 6; 18.84
35	Quartz or limestone outcropping	38; 46; 57.73	121; 6; 37.21
36	The Narrows	38; 47; 33.56	121; 6; 26.53

37, 38	Narrow's suspension bridge west abutment	38; 47; 35.10	121; 6; 23.62
39, 40, 41	Stone mason aqueduct	38; 47; 40.20	121; 6; 27.59
42	River cobble wall/landing	38; 48; 36.32	121; 6; 5.79
43	North Fork Ditch Horseshoe Bar	38; 48; 45	121; 6; 24
44	Riveted cast-iron pipe Horseshoe Bar	38; 48; 43.8	121; 6; 25.2
45	Native American grinding holes Rattlesnake Bar	38; 49; 5.81	121; 6; 6.94
46	Rattlesnake Bar Bridge northern abutment	38; 48; 52.92	121; 5; 29.39
47, 48	Rattlesnake Bar Bridge southern abutment	38; 48; 50.34	121; 5; 30.39
49	North Fork Ditch Mormon Bar weir outlet	38; 50; 6.6	121; 5; 22.79
50	North Fork Ditch retaining wall	38; 51; 53.39	121; 3; 21.59
51, 52	North Fork Ditch multi-weir concrete outlet	38; 52; 0.60	121; 3; 14.4
53, 54	Birdsall Dam eastern anchorage	38; 52; 40.27	121; 3; 24.98
55, 56	Knickerbocker waterfalls	38; 52; 21.54	121; 2; 47.13
57	Rattlesnake Bar from Wild Goose Flats	38; 48; 55.23	121; 5; 57.33
58	River gauging station	38; 48; 50.09	121; 5; 38.63
59	Wild Goose Flats mining operation	38; 48; 41.46	121; 5; 33.63
60	Native American grinding holes, Wild Goose Flats	38; 48; 14.13	121; 6; 0.52
Not Shown	Second Native American grinding holes	38; 48; 4.36	121; 5; 48.79
61	Stove parts	38; 47; 51.59	121; 5; 58.07
62	Small concrete foundation	38; 48; 1.69	121; 5; 51.86
63, 64, 65	Zantgraf Mine	38; 47; 45.21	121; 5; 45.19
66	Concrete utility pole piling	38; 47; 33.96	121; 6; 3.59
67	Peninsula campground concrete bridge	38; 45; 47.46	121; 6; 47.57
68	Native American grinding holes, Peninsula	38; 45; 16.81	121; 7; 3.63
69	Peninsula dam	38; 44; 27.55	121; 7; 45.19
70	Peninsula dam	38; 43; 48.39	121; 8; 23.23
71	Sunken boat	38; 44; 5.0	121; 8; 10.68
72	Massachusetts Flat mining operation	38; 43; 47.20	121; 8; 27.79
73	Massachusetts Flat rock hopper	38; 43; 47.33	121; 8; 28.78
74, 75	Massachusetts Flat ore cart and tailings	38; 43; 44.88	121; 8; 31.44
76, 77	Massachusetts Flat water ditch	38; 44; 17.32	121; 7; 52.22
78, 79	Peninsula gold mine coyote hole	38; 43; 2.98	121; 8; 15.65
80	Peninsula tip	38; 42; 57.90	121; 8; 26.39
81	End of the road, Peninsula	38; 42; 45.28	121; 7; 54.55
82	Concrete square shaft	38; 42; 49.51	121; 7; 55.05
83	Sunken fishing boat	38; 43; 15.01	121; 7; 58.89
84	Peninsula road concrete bridge	38; 43; 21.91	121; 7; 49.06
85, 86	Negro Hill building foundation	38; 43; 32.78	121; 7; 39.67
87	Mormon Island rock dam	38; 42; 0	121; 7; 58.29
88	Mormon Island cemetery	38; 42; 8.15	121; 7; 43.57
89	Mormon Island school	38; 42; 11.39	121; 7; 3.61
90, 91	Red Bank dairy	38; 42; 53.07	121; 6; 42.15

92	Red Bank rock foundation	38; 42; 53.33	121; 6; 40.42
93	Higgin's Point, west side	38; 45; 14.86	121; 4; 18.32
94	Higgin's Point, Salmon Falls	38; 45; 28.51	121; 4; 18.70
95	Higgin's Point house foundation	38; 45; 25.79	121; 4; 12.93
96	Negro Hill Ditch retaining wall	38; 45; 35.79	121; 4; 15.57
97	Rock mounds	38; 45; 33.59	121; 3; 59.04
98	Natomas Ditch concrete lining	38; 45; 19.98	121; 3; 42.94
99	Natomas Ditch rock retaining wall	38; 45; 14.79	121; 4; 0.59
100	Salmon Falls Bridge	38; 45; 37.15	121; 3; 41.92
101	Salmon Falls utility foundations	38; 45; 33.08	121; 3; 39.42
102, 103, 104	Negro Hill Ditch stone aqueduct	38; 46; 21.79	121; 2; 52.62
105, 106	Natomas Dam	38; 46; 22.89	121; 2; 23.18
107, 108	Natomas Dam fish and debris screen	38; 46; 21.60	121; 2; 34.77
109	Natomas Ditch hiking path	38; 46; 18.70	121; 2; 47.35

Sources

Additional sources of information not referenced in the text.

Gary Pitzer, *150 Years of Water: The History of the San Juan Water District* (Granite Bay, Calif.: Water Education Foundation, San Juan Water District, 2004).

California Digital Newspaper Collection, Center for Bibliographic Studies and Research, University of California, Riverside, http://cdnc.ucr.edu

John M. Letts, *California Illustrated by a Returned Californian, Including Description of the Panama and Nicaraguan Routes* (New York: William Mc Eldridge Publisher, 1852).

"Environmental Conditions, Geology, Folsom Lake State Recreation Area" (San Francisco: Geotechnical Consultants Inc., April 2013). "Environmental Conditions, Hydrology, Folsom Lake State Recreation Area, (San Francisco: Geotechnical Consultants Inc., Psomas. Sacramento, Calif., April 2003). "Folsom Dam Fact Sheet" (Sacramento, Calif.: U.S. Department of the Interior, Bureau of Reclamation, Mid-Pacific Region).

Thompson & West, *History of Placer County* Reprint (Oakland, Calif.: 1882)

John H. Plimpton Research Records. Multi-volume collection of research material on the Middle, North, and South Forks of the American Rivers by California State Parks Ranger John H. Plimpton, Placer County Archives.

History of a Place Called Rescue (Rescue, Calif.: Deer Valley Press, November 2011). "Historic Aerials," source of the 1952 aerial photographs included in this book, http://www.historicaerials.com/.

Map of the American River and Natoma Water and Mining Company's Canals, 1910. Center for Sacramento History, 1981/037/0821.

Sacramento, Placer and Nevada Railroad, August 30, 1861. Plan of the first to the lower division of the Sacramento, Placer and Nevada Railroad, Folsom to connection with the California Central Railroad. California State Railroad Museum Library. Tube 468 + D ID 33717

Author

Kevin Knauss is an independent health insurance agent living in Granite Bay, California. Most of his writing for his website, www.insuremekevin.com, focuses on health insurance, Obamacare, Covered California and Medicare. In addition to his hiking and history posts, he also maintains pages on his website devoted to railroad maps of Northern California, his mechanical clock collection, and posts relating to his small mid-century modern home.

www.ingramcontent.com/pod-product-compliance
Lightning Source LLC
Chambersburg PA
CBHW051338200326
41519CB00026B/7473